EUROPEAN M

Terms and conditions of employment

EUROPEAN MANAGEMENT GUIDES

General editor: Pete Burgess

Terms and conditions of employment

Incomes Data Services

Institute of Personnel Management

Phototypeset by Paragon Photoset, Aylesbury, Buckinghamshire
and printed in Great Britain by
Short Run Press Ltd, Exeter, Devon

British Library Cataloguing in Publication Data
Terms & conditions of employment. – (European management guides).
1. Europe. Personnel. Law
I. Incomes Data Services II. Institute of Personnel Management III. Series
344.0412

ISBN 0-85292-452-6

Contents

General Introduction

European Management Guides

The growing involvement of British businesses in the European Community has imposed novel demands on managers. For many small and medium-sized companies the challenge has been to find new markets and consolidate service networks. Larger organizations have been swept up in a series of mergers, acquisitions, joint ventures, strategic reviews and rationalization. In both cases, managers with personnel responsibilities have seen their tasks expanded and redefined. Increasingly, this includes involvement, both direct and at arm's length, in the recruitment and management of employees in other member states of the European Community. In many organizations the human resource manager has become the key figure in integrating the individuals and cultures which make up an internationalized company.

Experience has shown, often painfully, that the management of people lies at the heart of whether an international joint venture, merger or acquisition succeeds or fails. And familiarity with the local culture and regulation of employment underpins any understanding of the opportunities and limitations facing key executives in a foreign subsidiary. Acquisitions in particular, with their inheritance of established employment law and industrial relations practices, can pose a special challenge to new management teams.

The European Commission's plans for new Community legislation on many aspects of employment, as a part of the 'social dimension' to the single European market, will bring national systems of employment law closer together. However, differences between cultures, institutions, law and practice will persist for the foreseeable future. Understanding and working with the grain of this diversity will be vital for any personnel or line manager entrusted with European responsibilities. Professional advice is indispensable in any European venture, but precious time and money can be saved by gathering information in advance before meeting lawyers, consultants or the local public authorities.

European Management Guides aim to meet this need for accessible and comprehensive information on employment in the countries of the European Community. The series, researched and written by Incomes Data Services Ltd, and published by the Institute of Personnel Management, consists of five handbooks covering:

- Recruitment.
- Terms and conditions of employment.
- Industrial relations.
- Pay and benefits.
- Training and development.

Each handbook will present information on a country-by-country basis, and will be structured to allow easy cross-country comparison. Appendices detail local organizations which can provide further help and information. European Management Guides are not, however, intended as a substitute for expert advice, tailored to an individual situation and given as part of a professional relationship. Every effort has been made to ensure that the information contained in the handbooks is accurate and relevant. The publishers and authors offer them to readers on the understanding that neither organization seeks to take the place of either a lawyer or a consultant.

Incomes Data Services

Incomes Data Services has monitored employment developments in Europe since 1974. IDS's International Service publishes:

- *IDS European Report*, a monthly subscription journal on pay and employment law and practice in the European Community and Scandinavia. Each issue includes news on pay, collective bargaining and legal developments in Community countries, a Country Profile outlining recent developments, statistics and the economic background for individual countries, features, and regular statistics on pay, prices, labour costs and employment.
- *IDS International Documents*, two series of in-depth publications covering Commencement and Termination of Employment and Pay and Conditions on an individual country basis. Updated annually, each guide contains both context and detail for a single country in these central areas of human resource management.
- *1992: Personnel Management and the Single European Market*, published jointly with the Institute of Personnel Management.

For further details please contact: IDS Subscriptions, 193 St John Street, London EC1V 4LS (tel. 071-250-3434, fax 071-608-0949).

Acknowledgements

This handbook was researched and written by the International Department of Incomes Data Services Ltd, London. The series editor is Pete Burgess, and other contributions were prepared by Angela Bowring, Andrea Broughton, Nana Mallet, Sally Marullo, Tony Morgan and Raili Seppanen. The authors are grateful for the help and advice tendered by John Piperoglou (Stedima, Athens), Christina Valtazanou (Unilever Hellas), and Fernando Tempera. We also want to extend our gratitude to the panel of expert advisers drawn from the National Committees of the Institute of Personnel Management and to the IPM's Publishing Department.

Introduction

Terms and Conditions in Europe

This is the second handbook in the IDS/IPM five-volume series of European Management Guides. Following on from the first volume, which dealt with recruitment, it aims to outline the main statutory and agreed provisions on individual terms and conditions of employment in the European Community. The main areas covered on a country-by-country basis are:

- Contracts of employment, their formal requirements and the main rights and duties which arise out of them.
- Basic conditions, such as working time, time off, and retirement.
- Equality provisions.
- Individual termination of contract.

The book begins with an outline of existing and proposed European Community legislation on the areas set out above.

Collective arrangements are considered in this handbook only where agreements determine the terms and conditions of employment for individuals. Industrial relations, including statutory or agreed machinery for consultation and codetermination, the law on trade unions and industrial action are dealt with in volume 3 of the series. Volume 3 will also look at procedures for collective redundancy which under European Community law require an element of work-force consultation.

The determination of pay, national pay practices and trends will be dealt with in volume 4 of European Management Guides, which focuses on collective bargaining, pay and benefits.

Personnel practitioners in most countries in the European Community are accustomed to working in a more legally regulated framework for the determination of individual terms and conditions than is found in the United Kingdom. This framework typically consists firstly of a body of statute law, in some cases codified but often existing as a patchwork of individual employment laws together with sections from broader statutes, frequently very venerable, such as national civil codes. In contrast to UK practice, areas such as 'atypical' work, hours of work, leave and time off are customarily regulated by statute. And in many instances, the setting of these terms, as well as the form of the employment contract itself, may be subject to a statutory requirement to consult and obtain the consent of elected work-force representatives.

Statute law is complemented by a structure of collective agreements which typically in continental Europe have the force of law and frequently cover all companies in an industry or region whether or not they are signatories to the agreement or members of a signatory organization. Usually such agreements take precedence over the terms of statute law provided they offer more favourable provision to the employee. However, it is also sometimes the case that derogations from statute law are permitted on such issues as working hours arrangements or periods of notice only on the condition that any provision worse than that prescribed by law has been sanctioned by a collective agreement. (This principle is reflected in the proposed Directive on Working Hours discussed below.)

This more pronounced legalistic flavour has meant that much of this handbook is concerned with setting out the minimum statutory requirements which must be met and indicating how typical collective provision might build on them. The existence of a floor of legal and agreed provisions does not mean, however, that companies are denied the flexibility or scope to develop practices which best suit their operational, recruitment and retention needs. Firstly, these requirements are only a floor, not a ceiling. Secondly, legal regulation should not be confused with proscription, although it may entail an unaccustomed level of consultation with the authorities and work force. And thirdly, the development of legislation in continental Europe since the mid-1980s has been towards a form of controlled liberalization, in which the scope for flexibility on aspects such as working hours and contracts of employment has been enlarged, although — as in France, Germany and Portugal — often brought under a statutory regime for the first time.

European Community law

European Community legislation has so far had comparatively little direct impact on the determination of individual terms and conditions of employment, although the exceptions – principally equal pay and treatment and employee rights during a transfer of undertaking – have had a major impact on corporate pay and benefits policies, and have introduced new principles into the interpretation of Community law within the UK. (Issues connected with the application of EC provisions on the transfer of undertakings and collective dismissals in individual member states are dealt with in volume 3 of European Management Guides).

Equal pay and equal treatment

The impact of European Community law on equal pay and equal treatment has been felt in member states both through the national implementation of equality directives – providing for equal pay for work of equal value (75/117), equality of treatment (76/207), equality in social security (79/7) and occupational social security (86/378) — and a number of important decisions of the European Court of Justice (ECJ).

Under Article 119 of the Treaty of Rome, member states are required to apply the principle that men and women should receive equal pay for equal work, whether paid in cash or kind, or received directly or indirectly in respect of their employment. In 1976, the ECJ ruled that Article 119 could be applied directly in relation to direct and overt discrimination, and in 1987 that it could be used in circumstances relating to indirect discrimination (see below). The Equal Pay directive elaborates upon the principle of equal pay for work of equal value.

ECJ rulings in this area have compelled national governments to amend, or consider amending, national legislation which the Court has found to be incompatible either with existing directives or the equal pay provisions of the Treaty of Rome. The prime example in the UK is the Equal Pay (Amendment) Regulations 1983, which made equal pay claims possible for work of equal value, outside of an existing job evaluation scheme. Draft legislation has now been tabled in Federal Germany to respond to ECJ decisions on remedies for discrimination on appointment, and exclusion from sick pay under statutory provisions. The ruling by the ECJ in the Danfoss case (1989) has also now placed the burden of proof on the employer to show that a pay structure in which there is a statistical difference between men's and women's pay is not based on sex discrimination.

The fact that the majority of 'atypical' employees in most European countries — that is, people working on other than full-time indefinite contracts of employment — are women has also linked the issues of atypical employment and sex discrimination through claims against indirect sex discrimination brought under article 119 of the Treaty of Rome. The pioneer European case on this issue which was referred from a German court to the ECJ, *Bilka Kaufhaus v. Weber von Hartz*, was decided in 1987 when the ECJ ruled that an occupational pension scheme whose rules excluded part-timers indirectly discriminated against women on the grounds that more women than men were affected by the restriction, and that there were no objective factors to warrant it. ECJ law on indirect discrimination was also developed in 1989 with another German case, Rinner-Kühn *v.* FWW Spezial-Gebäudereinigung, where the court ruled that any national provision which excluded certain categories of part-timers contravened article 119, unless the exclusion was warranted by objective factors. In making its decision the ECJ rejected a submission from the German government that the employer's duty of care to part-time employees was reduced because such employees were 'less integrated in and bound to' the establishment. The ECJ has also required changes in collective agreements which have been found to be directly or indirectly discriminatory. For example, a 1991 ruling on the German public-sector pay agreement declared that differing qualifying periods for full- and part-timers for advancement in a pay progression system were indirectly discriminatory and had to be equalized.

Community influence to grow

However, the lack of binding prescription of individual terms will no longer apply should proposed Community legislation put forward on the basis of the principles set out in the 1989 Social Charter and associated Action Programme come into force. The drive for more Community legislation on terms and conditions of employment is a result of the decision to complement moves to create a unified market for goods, services, capital and labour in the EC by the end of 1992 with a 'social dimension' to ensure a minimum floor of social and employment standards. Although the Social Charter itself is non-binding, it has formed the basis of a number of proposed directives which would set a framework of minimum rights and principles which member-states would have to implement, either by amending national legislation, passing new primary legislation or incorporating into binding collective agreements. The passage of the 1987 Single European Act, which initiated the drive to the Single Market, not only marked a commitment to remove barriers to the free movement of goods, money and people, and introduce the required legislation, but also entailed a change in what had been a basic principle of the legislative process within the European Community.

Whereas, previously, legislation required the approval of all member states through the need for unanimity in the Council of Ministers, the ultimate EC decision-making body on new laws, the Single European Act introduced the notion of qualified majority voting. Now any legislative proposal deemed essential to complete the Single Market and introduced under article 100A of the Treaty of Rome (the amendment underpinning the Single Market project) can become law if fifty-four votes, of the total of seventy-six in the Council, support it: within the Council, each country has a number of votes weighted in line with the size of its population. Qualified majority voting was also extended to measures directed at health and safety and the working environment introduced under another Treaty of Rome amendment, article 118A. However, measures connected with taxation, freedom of movement of persons and the rights of employees still require unanimity: as has now happened with the passage of some draft directives through the EC's legislative process, the haziness around whether a particular measure concerns employee rights or their health and safety has created controversy, and could slow down the implementation of Single Market measures.

Qualified majority voting is also associated with an enhanced role for the European Parliament under a new 'cooperation procedure' between the Commission, the Parliament, and the Council of Ministers. This allows the Parliament to influence the legislative process on proposals made under articles 100A and 118A. Specifically, should Parliament reject a proposal which has passed through its initial consideration in the Council of Ministers and been approved by qualified majority voting, the Council may only finally approve the proposal unanimously when it returns. Furthermore, should Parliament amend a proposal, the Commission has one month to review it, and the Council of Ministers three months either to adopt any new Commission proposal by qualified majority voting or to accept Parliament's amendments or introduce its own unanimously. If the Council fails to act within the three months, then the initial proposal lapses.

As a response to criticism that draft directives were poorly prepared and insufficient attention was paid to the views of the social partners, the Commission has also introduced a pre-consultation stage under which organizations such as UNICE, representing employers at EC level, and the European Trades Union Congress can view the Commission's proposals and gather national opinion before they are published as fully fledged draft directives. The Commission also intends to review this machinery in 1991 and improve it where necessary.

At the time of writing, progress on Community social legislation under the Single Market programme has been slow, irrespective of whether proposals have required unanimity or may be presented through the qualified majority voting procedure. By Easter 1991 a 'common position' in the Council of Ministers – the first crucial step in the legislative process – had been arrived at on only one draft directive, that giving employees in 'atypical' employment equivalent health and safety protection to full-timers. Below we set out the main directives, as originally presented by the Commission except where stated, currently in the pipeline and proposed under the Social Charter which would have a direct bearing on individual terms and conditions of employment.

- *Proof of an employment relationship*. Under this proposal, submitted by the Commission on 5 December 1990, and scheduled for treatment under the unanimity procedure, all employees working eight or more hours a week must be provided with a 'written declaration' of their terms and conditions by the employer: details would have to include, the place of work, a job description and the category of employment, the duration of the employment and any probationary period, if appropriate, and periods of notice, working time, leave, pay, social security coverage, and which – if any – collective agreements terms may be derived from.
- *'Atypical' work*, defined as part-time, fixed-term and agency employment. There are three draft directives in this area, two covering atypical employment conditions, and one covering health and safety. Under the draft directive on employment conditions, presented on 13 June 1990, atypical workers would

have to be granted equal access to vocational training, equal access to company and non-contributory benefits, and priority consideration for full-time permanent vacancies. Atypical employees must also count as eligible employees in calculating the threshold at which employee representative bodies are established under national legislation, employers would have to consult work-force representatives on their use, and in larger companies would have to produce regular reports on their employment. Although initially tabled as requiring unanimity, the European Parliament has challenged this and the proposal has been returned to the Commission and Council of Ministers for decision.

A second draft directive on atypical employment deals with the issue of these employment relationships from the standpoint of competition between member states under the Single Market, and has therefore been tabled under qualified majority voting procedures (article 100A). The directive seeks to provide a set of minimum conditions for such employees, to prevent distortions of competition arising through disparities in rules governing atypical employment. It would require access on the same basis as full-timers to social benefits under statutory and occupational schemes for employees working at least eight hours a week, and pro rata entitlement to leave, severance pay and seniority allowances. Fixed-term contracts would not be allowed to exceed thirty-six months.

The health and safety draft directive, which would require a written contract for atypical employees, setting out the nature of their work, the qualifications required, and any hazards, with no temporary staff allowed on dangerous work, is subject to qualified majority voting.

- *Hours of work.* The draft directive on working hours, presented on 3 August 1990, has been tabled by the Commission as a measure under the health and safety provisions of article 118A, which allows for qualified majority voting. As originally presented it would provide for: a minimum daily rest period of eleven consecutive hours in each twenty-four-hour period; one rest day a week following on from the eleven-hour break, with the option of averaging this over a two-week period; an amount of annual paid leave 'in accordance with national practices'; controls on shift work and night work, with appropriate health and safety measures for night workers and those on rotating shifts. Following a continental European principle, derogations from these provisions could be agreed by collective agreement in the light of the circumstances of enterprises, provided the total amounts of compensatory rest are granted within a six-month reference period.

- *Pregnant women.* The draft directive on maternity protection, presented on 3 August 1990, is deemed to be a health and safety matter, and is subject to qualified majority voting. The directive, as it now stands after parliamentary amendment, would require a minimum period of paid maternity leave of sixteen weeks, with special protection for nursing mothers.

Draft directives proposed under the Social Charter and Single Market programme covering industrial relations issues, such as work-force consultation and information rights, will be dealt with in volume 3 of European Management Guides.

1

Belgium

The determination of terms and conditions in Belgium rests largely upon rights enshrined in the country's constitution, an extensive system of law, and binding nationwide and sectoral collective agreements. Although employment matters are generally decided at national level, there are some areas, such as training, which have recently come within the competence of regional authorities, which may issue their own decrees.

Contracts of employment

The terms of an individual contract of employment are governed, in order of priority, by:

- Legislation, including the Contracts of Employment Act (see below).
- Collective agreement. A nationwide collective agreement (*accord interprofessionnel*) concluded in the National Labour Council may be made statutory either totally or in part by royal decree. Collective agreements negotiated at industry/sector level (the backbone of Belgian industrial relations) within a bipartite joint committee (*commission paritaire*) on which employers and trade unions are equally represented, are binding on firms belonging to a signatory employers' organization but may apply to all firms in a particular industry if the agreement is extended by royal decree.
- Company agreements.
- The contract of employment.
- Internal works rules.
- Custom and practice.
- Verbal agreements.

The law on contracts of employment dated 3 July 1978 (*loi*

relative aux contrats de travail), with subsequent amendments, co-ordinates a number of legal texts on employment contracts for blue-collar workers and white-collar employees, as well as for commercial representatives, domestic workers and students, both in private industry and in the public sector. Home workers, occupational physicians and temporary workers are excluded from the scope of the legislation.

The law covers the type and form of employment contract, the obligations of the parties, and the suspension and termination of the contract.

Form and type of contract

There are significant differences laid down in law (and in collective agreements) between the contracts of blue-collar workers (*ouvriers*), white-collar staff (*travailleurs intellectuels*) and commercial representatives. Classifying occupations into these categories can be problematic. Differences in treatment under the law principally concern the length of probationary and notice periods, restrictive covenants, and arbitration clauses, which depend on the level of annual remuneration received. (Pay ceilings relevant to these terms are updated annually in line with the collectively agreed wage indexation system.) Holiday and sickness entitlement, which is principally governed by collective agreement, also varies as between blue and white-collar workers.

An employment contract may be agreed verbally or in writing, although certain types of contract must be written.

Contracts for a *fixed term* or for a specific assignment must be written, otherwise they will be deemed to be for an unspecified term. A series of fixed-term contracts, where there is no interruption in the work undertaken, will be considered equivalent to a contract of unspecified duration after a certain period, unless the nature of the work justifies the use of contracts for successive fixed terms.

A contract for a *part-time* worker must likewise be in writing. It must specify the total number of weekly hours and the agreed scheduling of those hours, or must make reference to appropriate documents (such as the collective agreement, or internal works rules) which give the information. Part-time work must amount to at least one-third of full-time hours worked in the firm, and each work period must last three hours or more. If flexible hours are to be worked, prior notice of the schedule of working hours must be given – usually in accordance with a sectoral agreement – or, failing that, at least five days' notice. The average weekly hours limit must be observed, calculated over a three-month reference period, or exceptionally over twelve months.

A contract of *replacement*, where a stand-in is engaged for reasons other than those associated with economic factors or industrial action, must be concluded in writing no later than the day employment commences for a maximum period of two years. Details of the reason for the replacement and of the employee being replaced must be provided.

Scope of a written contract of employment

A contract would normally outline the following:

- Identity of the parties and the premises where the work is to be carried out.
- Duration of the contract.
- Starting date, job title and duties.
- Working hours and schedules, including shifts, with reference where appropriate to other documents.
- Rate of pay; method and frequency of payment.
- Health and safety obligations, including issues specific to the job and/or the equipment to be used.
- Circumstances leading to suspension of the contract.
- Procedures for terminating the contract, whether by mutual consent, unilaterally, by summary dismissal or by *force majeure*. (*Force majeure* may not in itself be sufficient to terminate the contract but may temporarily suspend it.)
- Probationary period.
- Reference to works rules.

Implied contractual terms

In general, there must be mutual respect between the parties to a contract and propriety in carrying out its terms. Specifically, *employees* have a duty to:

- Carry out work with care, diligence and integrity under agreed conditions.

- Act under the employer's instruction.
- Refrain from divulging business secrets or competing unfairly during and after the term of the contract.
- Refrain from any action harmful to their own, others' or a third party's safety. (As in the UK, the employer would normally assume responsibility for damage or injury sustained by a third party — an employer could claim against the employee if fraudulence, misrepresentation, negligence or gross misconduct could be established.)
- Return to the employer any equipment or materials in good condition.

Employers have a duty to:

- Ensure the provision of work under the agreed terms and conditions, with appropriate equipment, materials and help being supplied.
- Provide health and safety care and protection, including first-aid facilities.
- Pay employees according to agreed terms. (Bonuses are deemed to be remuneration, and part-time workers must be paid pro rata to full-timers.)
- Permit employees time to carry out religious duties, and civic functions prescribed by law.
- Provide an induction programme, particularly for young people.
- Give existing part-time employees preference in filling full-time vacancies.
- Ensure equality of treatment.
- Provide facilities for the safe keeping of employees' own belongings and clothing.
- Complete appropriate social documents when a contract comes to an end.

Non-competition provision

A non-competition provision is at least an implied term of most contracts of employment, with the exception of blue-collar workers earning less than Bf 766,000 (£12,800) annually. Such a provision prevents an employee, whilst working for a particular employer and thereafter, from engaging in any activity, including communicating any commercial or industrial knowledge, which could prejudice that employer's business. For employees earning between Bf 766,000 and Bf 1,532,000 (£12,800–£25,600) annually the clause affects only those in certain jobs which have been identified in collective agreements concluded in the sectoral joint committees, or at company level. Such a provision is assumed to apply to staff earning in excess of Bf 1,532,000 (£25,600) a year unless they undertake functions exempted by collective agreement.

The situation is slightly different for commercial representatives earning over Bf 766,000 (£12,800) a year. The restriction can apply only to similar activities, to the territory which the representative was working in previously, and for a maximum period of twelve months.

There is no specific legislation governing *employee inventions*. The whole area of industrial property rights, including patents, trade marks, copyright and protection of designs, models and computer technology, has been the subject of discussion at various levels, spearheaded by moves by the European Commission.

Arbitration clause

Neither blue nor white-collar employees are generally permitted to agree to submit disputes arising from the employment contract to arbitration *before* the contract commences. However, an arbitration clause may be valid if it is contained in the employment contract of an employee whose annual pay exceeds Bf 1,532,000 (£25,600) or if the employee concerned is in charge of the day-to-day running of an enterprise or an establishment (*unité technique d'exploitation*).

Probationary period

Blue-collar workers may be subject to a probationary period, but only if it is stipulated on an individual basis and in writing. It may last between seven and fourteen days, but if the duration is not specified a seven-day period is assumed. Probationary periods of white-collar employees or commercial representatives may be no shorter than one month and may not exceed six months, for those earning less than Bf 920,000 (£15,370) annually, or twelve months for those earning more.

Claims arising from the contract

Claims arising from the employment contract are statute-barred one year after its termination, or five years after an incident which gives rise to a claim, whichever is earlier. The fact that a contract is deemed invalid because it contravenes employment legislation does not necessarily preclude an employee from lodging a claim on an issue arising from the contract.

Individuals' workplace rights

Individual grievances would normally be dealt with in the first instance by approaching the immediate line management and then by progressing the issue through the management hierarchy. If there is a trade union committee within the company, the individual employee has a right to be accompanied by a trade union representative when a grievance is being discussed. The trade union committee also has the right to be heard by management on an individual grievance when it has not been resolved through the normal channels. The works rules, to which every employee must have access, must state details of the employee representation bodies within the company.

Rights of access to information are mostly vested in the employee representation bodies, that is, the works council (*conseil d'entreprise*), obligatory in enterprises with over 100 employees; the health and safety committee (*comité de securité, d'hygiène et d'embellissement des lieux de travail*), obligatory in those with fifty employees; and the trade union committee.

Working Time

Hours of work

The principle legislation dealing with hours of work is the Employment Act of 16 March 1971 (*loi sur le travail*), as amended in 1978, establishing an eight-hour working day and a forty-hour week for establishments operating a six-day week. This law covers blue and white-collar workers employed in the private and in parts of the public sector. Other stipulations on working hours include a royal decree (dated 10 February 1965) on employees with senior responsibilities not subject to the law on hours of work, a royal decree No 225 of 7 December 1983 restricting overtime, the Economic Recovery Act 1985,the law of 17 March 1987 on the introduction of new schedules of working hours and collective agreement No. 42 of 2 June 1987 on consultation and negotiating procedures associated with introducing new schedules.

Weekly rest

Sunday must normally be allowed as a day off, except where employees are involved with the security of premises, cleaning, repair and maintenance work, accident prevention or other urgent work. There are exemptions which permit Sunday opening for some retail outlets (in the morning), spas and tourist resorts. Certain industries are authorized to work on twelve Sundays a year, but not more than four consecutively. Employees working on Sundays in these circumstances are entitled to compensatory time off within the subsequent six-day period.

Workers on continuous shifts may work Sundays provided they are granted an uninterrupted period of eighty consecutive hours off, once a week, eighteen of which must be on a Sunday (i.e. no more than six hours may be worked on Sunday).

There are further exemptions regarding Sunday, daily and weekly working hours which apply to those employed in some local authority services, in family businesses, transport and seafaring.

Daily and weekly hours

Under statute, working time must not normally exceed *eight hours daily* and *forty hours a week*. These limits may, however, be exceeded under certain circumstances.

The following exemptions do not require authorization from the labour authorities. Daily working time may increase to nine hours when employees are granted time off of half a day, a whole day or a day and a half *in addition to* Sunday, or to *ten hours* daily if the workplace is a long way from the worker's residence.

The Economic Recovery Act 1985 allows greater flexibility by collective agreement. Hours may diverge from the eight and forty-hour limits provided the agreement specifies:

- Average weekly working hours and total annual hours.

- The number of hours which may be worked in excess or short of the 'normal working day'. The deviation must not exceed two hours either way, with a nine-hour working day upper limit.
- The number of hours which may be worked in excess or short of the normal working week. The deviation must not exceed five hours weekly, with a forty-five-hour week upper limit. If the 'normal' working week in a firm is thirty-eight hours, then a collective agreement concluded under this legislation could set a working week ranging from thirty-three to forty-three hours.

The statutory limits may be exceeded under twenty-four-hour shift operations, where work is continuous by its nature, during stock-taking and during the preparation of annual accounts.

Authorization from the labour authorities is required if limits are to be exceeded in other cases, such as where the nature of the work means the limits on working hours are not practical; where materials are liable to deteriorate; transport services; repair and maintenance work which can be done only outside normal operating times; exceptional extra work loads; accident and urgent work.

Departures from the statutory limits are not granted if the working week, averaged out over a quarter (or, exceptionally, a year), would exceed forty hours.

In practice, most industries and sectors have established a *thirty-eight-hour week* or less by collective agreement, with day working hours falling between 06.00 and 20.00. The 1989/90 national framework agreement for white-collar employees not covered by any other sectoral collective agreement stipulated that by the end of 1990 all employees must work an average thirty-eight-hour week. The agreement covering white-collar staff in firms employing fewer than ten staff gives until the end of 1991 for a thirty-eight-hour week to be established, provided an extra payment amounting to one-thirty-eighth of wages is awarded in 1991.

Overtime

According to the royal decree No 225 which came into force on 1 January 1984, time off must be granted for overtime worked over a fixed period, normally during the quarter in which the overtime is worked. Under exceptional circumstances, up to sixty-five hours' overtime may be authorized, with time off granted in the following quarter. However, if overtime is undertaken in the context of an exceptional workload, accident, or urgent repairs, it may be compensated for by payment instead of time off.

The Economic Recovery Act of January 1985 allows for *two* types of exception to the granting of time off, provided an industry or company collective agreement exists. The *first* allows a departure from normal working hours, provided statutory or agreed limits are observed on an *annual* basis. The *second* extends the period during which compensatory time off may be granted from a quarter to a maximum of *one year*.

Flexible working hours

The law of 17 March 1987 on the introduction of new schedules of hours of work allows firms in the private sector to apply for exemption from protective legislation on working hours in order to use plant more efficiently and to have a positive effect on employment. Such exemptions cover Sunday working, work on public holidays, night work for men on technical, social or economic grounds, the extension of evening work for women and young people, the introduction of a twelve-hour working day and the removal of the forty-hour working week limit. Working hours schedules incorporating any of these exemptions must be introduced after consultation with the appropriate joint industry committee (*commission paritaire*), by an enterprise collective agreement endorsed by the on-site trade union delegation, if one exists, or by the joint committee itself. The nationwide collective agreement No. 42 of June 1987 gives details of the consultation procedures to be followed.

A recent collective agreement in the retail sector illustrates how this works. The industry is trying to streamline existing employment practices, for example by reducing overtime and the use of fixed-term contracts, whilst maintaining maximum flexibility. An employment contract drawn up for a five-day working week may now set a minimum weekly limit of eighteen hours, with a minimum three hours to be worked daily. The chain stores Sarma and Cora have set a limit of an average working week of twenty-four hours, with a lower limit of eighteen or an upper limit of thirty-six, with extra payments becoming due for more than four weeks at thirty-six hours.

Shift work and night work

In principle, women and young persons are prohibited from working between 20.00 and 06.00, although exceptions are permitted in certain industries and services.

As a result of the 1987 flexible working hours legislation, shift working has become more prevalent and working time practices generally more fragmented. In 1990 a nationwide collective agreement (No. 46) was concluded to create a framework for shift working, including night work, defined here as work between 23.00 and 06.00. It specifies that, as from 1 May 1990, night work may normally be introduced only through a collective agreement negotiated with enterprise trade unions. Night work can be done only on a voluntary basis and by employees on a permanent open-ended employment contract. Workers employed on night work have a right to transfer to day work after twenty years if aged 55 years or more, or at the age of 50 on medical grounds. Temporary transfer on to day work must be arranged for women during pregnancy and three months after confinement.

The first sectoral collective agreement encompassing night work for women was recently concluded for 20,000 white-collar workers in the import/export trade. This outlines details of the consultation procedure required before night work can be introduced. Night work has to be justified on the grounds of real economic necessity, and the proposal must be endorsed by on-site trade unions or, if they do not exist, by the joint committee itself. Proposals are then to be submitted to the Ministry of Labour, which can give authorization to go ahead on a six-month trial basis. At the end of the period the industry's joint committee must give its opinion. Account will be taken of issues such as the availability of transport. If the committee's view is favourable, then a collective agreement can be concluded. In principle, night work may be undertaken only on a voluntary basis.

Time off

Annual leave and public holidays

Full-time employees have a statutory right to four weeks' holiday a year. Further days of leave may be granted by collective agreement.

Employees do not always have a choice as to when holiday may be taken. An annual summer shut-down is common, and annual leave may be attached to public holidays or used to extend the Christmas and New Year break. A few industries grant extra holiday on the basis of long service and to young workers.

There are ten public holidays each year for which employees are entitled to be paid a full day's wage. If the holiday falls on Sunday, or another day not usually worked, employees are entitled to another day off, which may be fixed by collective agreement, by negotiation within a company or by mutual agreement. It is common for an additional day's holiday to be given when there is one normal working day between a weekend and a public holiday; the day is often referred to as a bridging day — a *pont*. In addition, 11 July and 27 September are commemorative days for the Flemish and French regions respectively, days which are not regarded as statutory holidays but which may be granted by collective agreement.

Maternity leave

The total length of statutory maternity leave is fourteen weeks: six weeks' ante-natal leave and eight weeks after confinement. In 1991 an additional one week's statutory leave, to be taken before confinement, was introduced. If an employee takes less than six weeks' leave before confinement she may add the amount not taken to the post-natal leave. Maternity benefit is payable from the state sickness fund.

Legitimate absence

This covers short periods of paid leave for personal reasons — a statutory right of all employees (royal decree of 3 December 1974) which may be improved upon by collective agreement. This includes family occasions, such as the marriage of the employee, for which two days are allowed, or that of a close relative (one day), a funeral, the ordination or entry into a convent of certain relatives, a child's first communion, the confinement of an employee's wife, the adoption of a child, military service induction, conscientious objection service, jury service and acting as polling station assessor at elections.

Paid time off may be taken by private-sector employees to fulfil political duties under a law of 19 July 1976.

Leave for urgent family reasons

The national framework agreement for the year 1975/76 allowed unpaid leave for urgent family reasons, a provision which has since been incorporated into some sectoral/industry collective agreements. In some cases the leave is paid. At the end of 1989 a national collective agreement (No. 45) granted up to ten days' unpaid leave for urgent reasons, which may include a child's illness, the hospitalization of a member of the family, damage to an employee's property or occasions when attendance is required by public authorities.

Career breaks

The Economic Recovery Act 1985 allowed a system of career breaks for either parent to become established in the private sector. A break lasting between six months and a year on a full-time basis may be taken by employees with six months' service if there is an industry or company agreement. (A part-time career break can last between six months and five years.) The 1989/90 national framework agreement required the social partners to consider establishing this as a right in the private (as it is in the public) sector, following which fifteen major industries have concluded their own sectoral agreements. If a vacated post is filled by an unemployed person, the employee on leave may receive an allowance of Bf 10,504 (£175) a month from the state. This rises to Bf 12,504 (£209) monthly if the leave is taken during the six years following the birth or adoption of a second child, and to Bf 14,504 (£242) for a third or subsequent child. Part-time workers are entitled to pro rata allowances. If the leave is taken immediately after the end of statutory post-natal leave, the government has granted a minimum twelve-week career break, instead of six months.

Educational leave

The scheme allowing employees to apply for paid time off to follow an educational course was originally established by law in 1973 but was amended in 1985 by the Economic Recovery Act. It is designed to encourage employees to take courses related to their employment *outside* working hours. For every fifty minutes of educational activity the employee is entitled to one hour off work, to an annual maximum of 240 hours (for closely linked vocational courses, or for a combination of vocational and general courses) or 160 hours (for the pursuit of more general courses). Pay is at the normal rate, and half is reimbursed to the employer by the Ministry of Labour. Employers of small companies have a right to refuse applications under certain circumstances.

Employee representation duties

A company may, depending on its size, have up to three different employee representation bodies: the trade union delegation, the works council and the health and safety committee. Employees sitting on these bodies are entitled to paid time off to carry out their duties but the amount depends on the size of the company, the number of bodies established and the provisions of the sectoral/industry agreement which covers the firm. The sectoral agreement for white-collar workers not covered by any specific industry agreement allows up to fourteen days' training for each office-holder on these three committees.

Military service

Young male employees may be called up for military service, which typically lasts eight to twelve months. The employer must preserve the employee's job during his absence and for a month after his return from service.

Retirement

A flexible retirement age between 60 and 65 years was introduced in the *state pension scheme* for male and female employees in the private and most of the public sector as from 1 January 1991. (Previously the normal statutory retirement age was 60 for women and 65 for men; men could opt for early retirement under the state scheme either on a reduced state pension or, under certain conditions, on an unreduced pension.)

A collectively agreed *early retirement* regime was introduced in 1974 under which schemes are set up on an industry basis by collective agreement and provide for employers to pay a supplement to older male workers who are made redundant from the age of 60 until they draw their state retirement pension, formerly at 65. Qualifying employees are required to have completed ten years' service in the same industry and a total of twenty years' employment. The original scheme has been amended several times, and some sectors provide for early retirement from as early as 50 years of age, though the usual lower limit is 58.

In 1990 the social partners concluded an agreement to restrict access to agreed early retirement pensions; the agreement has been made statutory (with some transitional provisions) and will affect employees retiring after 1 January 1991. In future the lower limit in schemes may not be less than 58 years; qualifying employees must have twenty-five years in salaried employment, with those over 60 required to have ten years' service in the industry over the previous fifteen and a total of twenty years' service.

Equal treatment

Discrimination on grounds of sex

Article 6 of the constitution guarantees equal treatment between men and women as regards access to employment and the professions, training and retraining and terms and conditions of employment. This principle was transposed into legislation through the economic reorientation law of 4 August 1978 (Part V, articles 116–53).

Equal treatment is defined as the absence of direct or indirect discrimination on the grounds of sex, including by reference to marital or family status; special provisions to protect maternity are permissible. Equally, measures promoting opportunities for women in employment and training, particularly to overcome existing inequalities, are also allowed (see below).

The first volume of this series of European Management Guides, on recruitment, outlines how equal treatment must be observed in recruitment and selection procedures. It must also be maintained as regards contractual terms and conditions of employment (including apprenticeships), collective agreements, promotion, employee representation in the workplace, pay (governed by a national collective agreement of 15 October 1975, No. 25, extended by royal decree) and collective dismissals, where criteria for selection must be set on a non-discriminatory basis.

Complaints of sex discrimination may be taken up, on a member's behalf, by trade unions or employers' organizations covered by the law of 5 December 1968 on collective agreements, by professional organizations representing the public sector and by bodies representing the self-employed. The burden of proof in establishing non-discriminatory conduct falls on employers.

An employee who has lodged a complaint of discrimination — under company procedures, through the labour inspectorate or before a labour court — may not be dismissed for reasons associated with the claim or have his or her contract unilaterally altered. Should it happen, the complainant has the right to ask the labour court for reinstatement.

Applicants who win their case in a labour court are entitled, at their choice, either to six months' pay or damages actually suffered, whether they have asked for reinstatement or not. Criminal penalties may also be imposed on those infringing equal treatment legislation (once or repeatedly), or on parties hindering investigations into alleged breaches. The penalties range from fines for individual offences of Bf 26–500 (40p–£8) and/or from eight days' to one month's imprisonment. The legislative programme of 1989 strengthened measures for monitoring the implementation of the equal treatment principle.

A royal decree of 14 July 1987 aims at promoting equal opportunities in the private sector. Positive action plans which either remedy existing inequalities between women and men or improve women's position and participation may be introduced either at sectoral level or within companies. The plans must specify:

- The objectives in terms of changing the comparative situation of men and women.
- A description of the positive action to be implemented.
- The plan's date of commencement and the dates fixed for assessing progress.

Regular appraisals must be carried out at the appropriate level within joint bodies — such as industry/sector committees and enterprise committees. The Ministries concerned with employment, social security and social affairs are charged with overseeing progress in these matters.

Sexual harassment

Centrally, this issue has been considered by the bipartite consultation body, the National Labour Council. In its most recent statement the council took the view that existing legislation was adequate to deal with incidents of this type. It also appealed to public bodies to make employees, particularly the young, more aware of the problem and to disseminate information within places of work.

Racism

Legislation passed in July 1981 forbids anyone who supplies goods and services in a public manner from discriminating on grounds of race, colour, ethnicity or nationality. Fines for infringement range from Bf 25 to Bf 200 (40p–£3.25) and/or eight days' to three months' imprisonment. Employers can be found responsible if they instruct employees to behave in a discriminatory manner.

The disabled

Legislation passed in 1963 places a general obligation on companies employing more than twenty staff, in the private or the public sector, to hire a quota of handicapped workers — defined as those with up to 30 per cent physical or 20 per cent mental impairment. However, this does not constitute a statutory requirement, since no implementing decrees have been issued. Around fifty of the sectoral joint committees have examined the issue and in some cases have issued policy statements.

The official placement services have a duty to find employment for handicapped people who are registered with them having initially been registered with National Fund for the Social Reclassification of the Handicapped (*Fond national de reclassement social des handicapés*). Subsidies are available to employers who hire disabled workers in respect of the first year's pay, adaptation of the work station, and to employees themselves in the form of clothing and equipment grants and help with travel costs.

Young persons

Currently young people who are in full-time compulsory education, which lasts at least until the age of 15 years, are not allowed to work. There are some exceptions, such as those engaged as actors, models, entertainers, etc. Young people under 18 may not work nights, undertake overtime or be employed on Sundays or public holidays, nor may they work more than a ten-hour day or longer than four and a half hours at a stretch without a break. Draft legislation is in preparation to control more strictly the employment of children, young people and students. The proposals include a complete ban on employing young people aged 18 or under who are in full or part-time education, and the provision of a written contract stating hours and setting a maximum daily work limit of eight hours.

Work environment

The general obligations incumbent upon employers have already been mentioned above in the context of the contract of employment. More specifically, employers have to ensure the protection of employees at work under the provisions of title 2 of the General Regulations for the Protection of Employment, laying down detailed requirements in areas such as the use of machinery, the handling of materials and fire risks. They must also comply with regulations on the hygiene of the premises, the environment, temperature, ventilation, heating, sanitation, provision of washroom facilities, toilets, supply of drinking water and canteens.

Regulations govern the provision of transport, special work clothes and safe keeping for employees' own belongings.

First-aid facilities, including first-aid boxes, must be readily accessible to ensure the prompt rescue and treatment of an injured party and their transportation to hospital. A sufficient number of people able to offer first aid must be trained. *Firefighting equipment* must also be available.

In all workplaces, irrespective of size, the employer must nominate a *safety officer* (it may be the employer if the work force totals fewer than twenty), together with assistants as necessary. The designated officer must have an adequate knowledge of health and safety legislation. The nominee may be advised by safety experts from both within and outside the firm. A safety officer is required to:

- Co-operate with routine safety checks (carried out by authorized private firms in Belgium), and inspections required by the labour inspectorate, for example following an incident.
- Remedy faults brought to their attention and advise on the renewal and installation of equipment, materials, etc.
- Advise the company health and safety committee.
- Arrange for the information and training of the work force and particularly of operatives and first-line management.
- Give the employer regular written safety reports on a monthly and annual basis, and following any specific incident.

Workplace *health and safety committees*, on which elected employee representatives sit, must be constituted in firms employing over fifty people, or in firms with at least twenty workers if such a committee existed at the time of the previous election (held at four-yearly intervals). The size of the employee representation depends on the firm's size. The employer side comprises the employer or appointed deputy, the designated safety officer and one other representative. The committee must meet at least once a month. Its members are given time off at normal rates of pay for undertaking health and safety committee duties. The committee's remit is:

- To examine reports produced by the health and safety department.
- To assist official Ministry of Labour checks and investigations.
- To collaborate with the employer in instituting and implementing a prevention policy.
- To make recommendations on health and safety matters.
- To seek to improve the appearance of the workplace and its surroundings.

Employers must set up a *medical service*, either on their own account or on the basis of facilities shared with other organizations. Certain recruits are obliged to undergo a medical, including all workers under 21 years of age, those exposed to risks or working with machinery and vehicles, and registered disabled. In addition the employer must warn all workers who may be exposed to certain risks (e.g. connected with gas escapes, fire, explosions). Medical records must be kept of all employees under surveillance.

Legislation was passed in 1987 banning *smoking*, at least in all communal areas of public buildings, including schools, hospitals, homes for the elderly and places of leisure and entertainment. Outside this, policy on smoking is most likely to be determined on an establishment basis in compliance with existing health and safety regulations.

Individual termination of contract

Employment legislation is weighted in favour of protecting an employee's job. Therefore employers must follow a number of procedures in connection with contract termination and, in appropriate cases, periods of notice (both statutory and agreed) must be observed or payment made in lieu; in the event of summary dismissal, reasons must be given.

No distinction is drawn between individual dismissal and redundancy. Normally no consultation is required. Some employees enjoy special protection against dismissal, such as employee representatives, pregnant employees, those on military service, those working with toxic waste. Employers are not normally bound to give reasons for the dismissal of an individual on an open-ended contract unless a summary dismissal is involved. Compensation is payable if the correct notice is not given.

Suspension of the contract

The contract of employment is suspended in a number of circumstances. They include: maternity leave, taking time off for reasons defined under legal or agreed provisions (for training), for the assumption of political office, military obligations, an employee's illness or an accident. The employer who wishes to terminate a suspended contract must give a reason and notice in accordance with legal and agreed provisions. Until 1990 the employer could, on payment, terminate the contract of an employee who had been off sick for six months and was unable to return to work. However, it is now incumbent upon employers to try to adapt the work to a returning employee.

Grounds for termination

A contract may be terminated for the following reasons:

- In the case of a fixed-term contract, on expiry of the term or on completion of the defined task.
- By mutual consent, by either party giving notice or, without notice, for a serious breach of contract.
- Retirement: for those reaching retirement age, notice to terminate the contract must be given, but the period is shorter than for other dismissals.
- Some instances of *force majeure* where it is impossible to fulfil the contract.
- Operational reasons.

Termination with notice

No obligations are laid down in law when the contract is terminated by mutual consent, though the contract of employment itself may state the procedure to be followed.

The employer must notify the employee of the termination of the contract by means of a registered letter or a writ from the clerk of a court. An employee may give notice by registered letter or by written notice handed over personally. Notice for a manual worker takes effect on the Monday following the week in which notification was given, and for a white-collar worker in the month following notification.

An employer who fails to give proper notice becomes liable to pay compensation equal to the period of notice.

Period of notice

Under the Contracts of Employment Act 1978 (articles 59, 60) employers must give *blue-collar employees* a minimum period of notice as follows:

- Not less than seven days, for workers with less than six months' service, and twenty-eight days for those with up to twenty years' service.
- Fifty-six days for those with twenty or more years' service.

Blue-collar workers must give employers fourteen days' notice after six months' service or twenty-eight days' after twenty years' service or more.

Under collective agreements, employers often have to give more generous notice. The iron and steel industry agreements specify that blue-collar workers must be given twenty-eight days' notice if they have less than ten years' service, fifty-six days (between ten and twenty years) and 112 days (over twenty years). If the employee terminates the contract, they may give shorter notice than under statute: fourteen days, with less than ten years' service, twenty-one days (between ten and twenty years) and twenty-eight days (over twenty years).

The Contracts of Employment Act (articles 82 and 83) specifies that *white-collar employees* paid less than Bf 766,000 (£12,800) p.a. are entitled to:

- At least three months' notice for less than five years' service.
- Six months between five and ten years' service.
- Nine months between ten and fifteen years' service.
- Twelve months between fifteen and twenty years' service.
- Fifteen months between twenty and twenty-five years' service, with a further three months for each additional five years.

These legal minima apply to all white-collar employees, but for those (the majority) earning more than Bf 766,000 (£12,800) the actual period of notice is subject to negotiation between the company and individual employee, once the employee has been informed of an impending dismissal. It must be no less than four and a half months for those earning Bf 766,000 to Bf 1,532,000 p.a. (£12,800–£25,600) or six months for those earning more.

One of a number of commonly used formulas is usually adopted to determine length of notice, normally well in excess of the legal minima. The two most popular are the Claeys formula and the Major scale. The main determinants of the length of notice are length of service, level of responsibility, age and salary. If no agreement is reached, the employer may impose notice, which can be challenged by the employee in the labour courts. In a recent decision by a Brussels labour court a period of notice of twenty-one months was fixed for a surveyor aged 37 with over twenty-one years' service and earning Bf 841,608 (£14,100) in 1987.

For *part-time white-collar* workers the length of notice is determined by earnings.

If a white-collar employee or commercial representative has found another job, they only have to give the employer:

- One month's notice if earning less than Bf 766,000 (£12,800).
- Two months if receiving Bf 755,000 to Bf 1,532,000 (£12,800 to £25,600).
- Up to a maximum of four months by agreement when notice is given if annual pay exceeds Bf 1,532,000.

During the period of notice, when the contract is terminated by the employer, an employee may take leave with pay for job-hunting of up to one day per week in the case of those on less than Bf 766,000. For those on higher salaries, leave amounts to one day a week for the last six months' notice but only half a day a week during the period of notice prior to that.

In Belgium the contract is not automatically terminated when an employee reaches retirement age. Notice must be given, although it is shorter than in the above circumstances. It amounts to six months if given by the employer or three months if given by the employee, reduced by half if the period of employment has been less than five years. Since the introduction of a flexible retirement age of between 60 and 65 years (see above) such notice may not be given by the employer until the employee reaches the age of 65 — that is, the upper age limit under the state retirement scheme.

Summary dismissal

The employment contract may be terminated without notice or payment in lieu for a serious reason (*motif grave*) involving 'any conduct which immediately and permanently precludes the continuation of a working relationship between the employer and employee', to use the wording of the Contracts of Employment Act 1978. This would include theft or persistent unauthorized absence by the employee. The injured party must take action to terminate the contact within three days of the serious incident (or latest in a series of incidents). Methods of notifying the termination are as for termination with notice. The injured party must establish the reason for the termination and communicate it to the other party in the notice.

Appeals

In principle a dismissal is deemed to be unfair if the reasons for it are unconnected with the individual's performance or with the requirements of the employer's business. Appeals against dismissal may be taken to a labour court, over which a qualified judge presides with the assistance of an employer and an employee representative. Every district has a labour court but there are only five courts of appeal, located in Brussels, Antwerp, Mons, Ghent and Liège. It is for a labour court to decide whether the conduct which led to the dismissal was of a sufficiently serious nature as to justify taking such action. Although internal works rules often give examples of behaviour which constitutes gross misconduct, a court would also look at other associated circumstances in the light of case law.

Employees are not normally entitled to reinstatement in their job in the case of successfully contested dismissals but are awarded compensation equivalent to (at least) six months' notice.

Procedures on termination of contract

Employers must keep an up-to-date register of all the personnel they employ. The register contains each individual's personal details and information on wage payments made. In addition, it holds data on schemes to which the employer belongs for the purposes of holiday payments, accident insurance, family allowance allocations, the sectoral joint committee to which the enterprise is attached and the organization which deals with the firm's payroll and social and administrative affairs (e.g. payment of social security contributions). The employer also has to compile an individual dossier containing personal details, information on wage payments, dates of the start and termination of the contract. Once a contract comes to an end, this document must be updated within two months and given to the employee who has left.

Appendix

National Employment Office
boulevard de l'Empereur 7
1000 Brussels
tel. +32 2 510 20 11
(French-speakers)
+32 2 513 89 42
(Flemish-speakers)

Ministry of Labour
rue Belliard 53
1040 Brussels
tel. +32 2 230 90 10

Sécrétariat de la Commission du Travail des femmes (information on positive action for women)
rue Belliard 51
1040 Brussels
tel. +32 2 233 40 16

Collective Industrial Relations Service (for copies of industry/sectoral collective agreements)
rue Belliard 51
1040 Brussels
tel. +32 2 233 41 49

Conseil National du Travail (for copies of national collective agreements)
avenue de la Joyeuse Entrée 17–21
1040 Brussels
tel. +32 2 233 88 11

Fédération des Entreprises de Belgique
rue Ravenstein 4
1000 Brussels
tel. +32 2 515 08 11

Ascobel (Association Belge des Conseils en Organisation et Gestion)
avenue Louise 430/12
Brussels 1050
tel. +32 2 648 10 55
fax. +32 2 646 05 41

2

Denmark

The 1953 Danish constitution only marginally affects the rights and obligations of employers and their workers. However, article 78.1 together with article 79 — which gives all citizens the right to form associations for any lawful purpose and the right of assembly — is the foundation for the legal operation of trade unions. Other provisions of the constitution which might appear to constitute a considerable protection, such as article 74, prohibiting any restraints on 'free and equal access to trade', and article 75.1, which gives citizens the right to a 'secure existence', have been interpreted by the courts to be only laudable principles.

Contracts of employment

In the broadest sense a contract of employment is an agreement which lays down the terms and conditions on which an employee agrees to sell their labour. Denmark has no general statute on employment covering all groups of employees, but, rather, separate legal provisions covering specific categories. The main statutes are:

- Salaried Employees Act (*Funktionærloven*)
- Vocational Training Act (*Lov om Erhvervsuddannelser*)
- Agricultural Workers and Domestic Servants Act (*Medhjælperloven*)

In addition most of these employment relationships are subject to collective agreements. However, in the case of blue-collar workers, basic terms are almost exclusively provided for by collective agreements. Whether a certain statute or collective agreement applies to a particular contract depends on the employee's occupation.

Other relevant statutes of general application include:

- Contracts Act (*Aftaleloven*), which lays down basic legal principles.
- Holiday Act (*Ferieloven*), which guarantees employees the right to paid annual leave.
- Supplementary Old Age Pensions Act (*ATP loven*), which obliges both employee and employer to contribute to a pension fund in the proportion of one-third and two-thirds respectively.

In Danish law, an agreement is valid whether it is oral, written or implied, although in practice almost all commercial contracts are written. A written contract of employment is advisable, in anticipation of disagreements which might arise later, especially if the parties wish to agree a special term not normally provided for under statute or by collective agreement. In most cases it will be evident which legislation and/or collective agreements apply as minimum requirements in any particular employment relation, with the employers free to offer better provision if they wish.

A variety of major labour organizations such as the Danish Employers' Confederation (DA) and the Clerical and Office Workers' Union (HK) have prepared standard contracts covering different areas of employment. The use of these standard contracts is widespread and has the advantage of speed and simplicity in employment arrangements. Some of the most basic conditions of employment which should be contained in a contract of employment include:

- The scope of the work.
- The duration of the agreement if it covers temporary employment.
- Pay.
- Working hours.
- Holiday periods.
- Sickness.
- Issues such as confidentiality.

Types of employee

There are four main categories of employee:

- Senior manager (*direktør*).
- White-collar (salaried) employee (*funktionær*).
- Blue-collar (wage-earning) employees (*lønmodtagere*)
- Vocational trainees and apprentices (*lærlinge og EFG elever*).

Senior managers comprise a special category of employee, defined by the exercise of a high degree of autonomy in their employment. As a consequence they are not deemed to fall within the scope of legislation and agreements covering contracts of service. In the absence of such general regulations, contracts of employment for senior managers must contain all the details of the employment relationship.

White-collar employees, defined in section 1 of the Salaried Employees Act, constitute one of the largest groups of workers in Denmark. The Act covers:

- Clerical and office staff.
- Technical or clinical staff.
- Persons whose work consists mainly of the supervision of other employees on the employer's behalf.

The Act applies only to employees working for at least fifteen hours a week and occupying a position under the supervision and direction of an employer. Any employee who falls within the scope of the Act is automatically granted various rights, such as a certain period of notice, dependent on seniority, and a right to full pay during sickness (see below). However, the Act is not the sole source of the respective rights and duties of employer and employee. For example, pay is regulated by collective agreement.

The rights and duties of *blue-collar employees* are almost exclusively laid down in collective agreements. The employment relationship may be based on a contract of employment which simply refers to the rules laid down in the relevant collective agreement, but more often employees are engaged on the basis of an oral agreement with the employer. The Danish Confederation of Trade Unions is attempting to change this practice so that every worker obtains the right to a written contract of employment.

Vocational trainees and apprentices. Before concluding an apprenticeship or traineeship contract the employer needs to obtain general official approval. Any training agreement not endorsed by the local job centre is invalid.

Probationary period

In the case of white-collar workers, the Salaried Employees Act provides for a probationary period of up to three months. During this period the employer is required to give notice of at least fourteen days, while the employee is free to terminate employment without notice unless the two parties have agreed otherwise. The period of notice for blue-collar employees is usually short, and probationary periods are therefore a rarity. As from January 1991 the probationary period for formal apprenticeships and trainees is three months.

Restrictive covenants

As part of the duty of loyalty to the employer an employee must not pass on confidential information to a third party. Section 9 of the Marketing Act forbids employees to pass on trade secrets without the employer's authority. This prohibition lasts for a period of three years after the termination of the contract of employment. Beyond this a contract of employment may include a clause about trade secrets and a non-competition clause which prohibits the employee from engaging in any activities directly in competition with the employer. Under the Salaried Employees Act such a clause is valid only if entered into with a salaried employee who holds a position of trust. The non-competition clause can remain in force only one year from termination of the employment unless the employee is compensated fairly for the continuation of the arrangement. This compensation must be fixed in a written agreement and appear as a distinct sum. If a non-competition clause is found to restrict the opportunities for the employee to exercise their occupation in a manner disproportionate to the compensation the courts may declare the clause invalid. More generally the Contracts Act (section 38) allows a non-competition clause to be wholly or partly disregarded or rendered unenforceable if its contents would be unfair or contrary to proper practice.

Employee inventions

In principle the copyright in an invention belongs to its creator; however, if the product is the direct result of a task assigned by an employer to an employee the copyright will normally belong to the employer. In order to prevent any disagreements arising later it should be made clear in the contract of employment who can lay claim to the copyright. In the absence of any ruling in the contract of employment or any implied agreement the Employee Inventions Act contains supplementary provisions. Under section 5 of the Act the employee is entitled to the copyright in the invention but the employer has a claim to its utilization if the invention was made in the course of duties at the workplace and falls within the field of the company's activities. The employee may additionally be entitled to lodge a claim for compensation.

Individuals' workplace rights

Employee grievances

Grievances raised by individual employees are usually dealt with first through the mediation of shop stewards, working through workplace co-operation machinery (see volume 3 in the European Management Guides series, covering industrial relations). However, there is a fundamental distinction between claims that arise from collective agreements and those that arise from an individual agreement between employees and employers. In the first case an employee may turn to a trade union for legal assistance; if mediation remains unsuccessful and the trade union supports the employee's claim it will forward the case to the labour court and act on the employee's behalf. In the latter case, or if the employee is not a member of a trade union, the employee may individually bring a claim before court.

Under co-operation agreements concluded between trade unions and employers' associations, the most significant of which is that between the Danish Federation of Trade Unions and the Danish Employers' Confederation, companies with at least thirty-five staff may establish a co-operation committee if either any employee or the employer wishes to do so. The committee's task, among others, is to ensure that the individual employee is informed about workplace issues through their immediate superior.

Data protection

Private employers are covered by the Registration Act 1987, which prohibits the collection of certain information by the employer. The Act provides that:

- Information about an individual's purely private affairs — such as race, religion, political affiliation, criminal offences — must normally not be held on file.
- Co-ordination of enterprises' different databases on employees is prohibited.
- Registered persons have a legal right to information about themselves which may be in a private database.

In principle a private database may contain only information on employees which is necessary for the enterprise, such as name, address, qualifications, remuneration, tax and periods of sickness. Violations of the Act may be prosecuted by the Registration Supervisory Board (*Registertilsynet*).

Working time

Hours of work

Provisions on working hours are laid down in collective agreements, in legislation and sometimes in the individual contract of employment. There is no general statutory provision on maximum working hours, with the exception of the Health and Safety at Work Act, which specifies breaks. Following the 1987 bargaining round, a progressive reduction in the working week to thirty-seven hours for full-time employees as from March 1991 was agreed for the vast majority of employees. Most employees usually have an additional half-hour lunch break every day. In principle the employer is also entitled to determine when periods of rest should take place, although in reality this right is limited by custom and practice. Section 50 of the Health and Safety at Work Act provides that working hours must be arranged so that the employee has a continuous period of rest of at least eleven hours within each twenty-four-hour period, calculated from the start of the normal working day. In addition, employees are entitled to twenty-four hours off work within each period of seven consecutive days.

Overtime and night work

Any work beyond the collectively agreed normal working week constitutes overtime and is normally remunerated according to schemes of payment set out in collective agreements. The amount of supplementary payment (*tillæg*) varies considerably as between agreements. In many cases an agreement will indicate the hourly amount to be paid as overtime as a sum of money. In other cases the supplement may be indicated by a percentage of the hourly rate, as is the case in the collective agreement between the Danish Employers' Confederation and the Office and Shop Employers' Association on the one hand and the Trade and Office Workers' Union on the other. Under this agreement, overtime must be paid at time-and-a-half for the first three hours, with double time for any subsequent hours, as well as for overtime on Sundays and public holidays. This agreement — as with many others — also allows employees to take time off in substitution for overtime pay. Overtime pay is frequently used, especially in areas where the collective agreement is based on a minimum wage.

White-collar employees may be required to work 'reasonable' overtime, provided notice is given.

There are no special provisions on night work, except that young people are not allowed to work between 20.00 and 06.00. Most collective agreements provide for double time to be paid.

Time off

Annual leave

Annual leave is regulated by the Holidays Act, which covers all employees in the public and private sectors except domestic servants and agricultural workers, or employees covered by a scheme offering terms at least as generous as those laid down in the Act. Under the Act all employees are entitled to five weeks' annual paid leave. The right to such leave within a working year accrues according to time worked: each month of employment entitles the employee to two and a half days' holiday with pay. Leave accumulated during the calendar year must be taken in the period beginning 2 May that year and extending to 2 May the following calendar year. Employees may take three consecutive weeks' leave during the period 2 May to 30 September. All employees receive a holiday bonus.

Public holidays

There are ten public holidays, of which one, Constitution Day (5 June), is only a half-day by statute.

Sick leave

Sick leave and sick pay are regulated by the Daily Sickness Benefit Act 1989, which came into force on 2 April 1990. It provides for sickness benefit for all categories of employees, payable by the employer for the first two weeks of sickness. Levels of payment depend on the category of employee. For white-collar workers the Salaried Employees Act provides that salaried employees are entitled to receive full pay during sickness. This is wholly paid by the employer for the first two weeks, but the employer may claim back up to Dkr 2,457 (£219) per week from the local authority thereafter. In contrast, blue-collar employees are not entitled to full pay during sickness unless provided for in a collective agreement. They receive sickness benefit payable by the employer for the first two weeks equivalent to earnings up to the current maximum hourly rate of Dkr 66.41 (£5.95); thereafter, responsibility for payment passes to the local authority. The maximum length of entitlement to sickness benefit is fifty-two weeks.

Maternity leave

Basic provisions relating to maternity leave are laid down in legislation, although supplementary provisions may be found in collective agreements. Under section 7 of the Maternity Leave Act a woman has the right to a period of four weeks' leave before the expected date of confinement and fourteen weeks' after the actual date of birth. In 1984 fathers gained the right to paternity leave, consisting of two weeks' leave after the child's birth. In addition parents have a further right to ten weeks' leave after the mother's fourteen weeks' leave has expired. This ten weeks' leave may be used by either the father or the mother, but only by one of them at a time. Someone who adopts a child is allowed twenty-four weeks' leave, provided the authorities require the parent to be at home when the child joins the household.

Payment for maternity leave is as with sickness benefit. Blue-collar workers are entitled to the statutory benefit level and statutory leave. White-collar employees, however, are entitled to leave on half-pay for up to five months, with employers refunded for part of this period at a maximum rate of Dkr 2,457 (£219) per week. The leave must be spread so that either three months are taken before, or three months are taken after, the date of confinement. Fathers may claim benefit from the state system during their period of leave. Mothers must give eight weeks' notice of their intention to return to work.

These provisions may be improved on, and are, in collective agreements. Typical provisions give employees full pay for part of the eighteen-week leave, usually for fourteen weeks.

Other reasons

One day's leave for *family events*, such as marriage, a move, or the death of a close relative, is generally granted, and may be regulated by collective agreement.

In most collective agreements it is presumed that the duties of a *trade union workplace representative* should affect neither working hours nor production. However, in bigger companies shop stewards may be released from normal work. Under the Health and Safety at Work Act *safety representatives* are entitled to time off for their duties.

Retirement

The social security system provides for a flat-rate old age pension (*folkepension*) from the age of 67; for a full pension, the person must have forty years' permanent residence in Denmark. The 1991 rate for married persons is Dkr 53,856 (£4,817) and for unmarried people Dkr 55,956 (£5,005). Moreover, retired employees are entitled to a supplement linked to lifetime earnings from the Labour Market Supplementary Fund (ATP) which amounts to Dkr 7,224 (£645) annually at maximum seniority. Employees aged between 60 and 67 who have been members of an approved unemployment fund for ten years within the previous fifteen years are eligible to retire early on an earnings-related benefit: the payment is 90 per cent of earnings for the first three years, and 80 per cent until the age of 67 is reached. In addition to the state pension, employees may join private supplementary pension schemes. Some unions have sought to negotiate additional supplementary pension schemes, and so far these have been accepted by employees in the public sector.

The basic state system is financed through general taxation, following the reform of social insurance contributions and their replacement largely by VAT. The ATP supplementary pension is financed by flat-rate employee/employer contributions.

Equal treatment

Sex discrimination

Generally, employers in the private and public sectors may appoint staff according to their judgement. However, this free choice is restricted by the Equal Opportunities Act, which prohibits discrimination on grounds of gender such as unequal treatment due to pregnancy, or matrimonial or family status. According to the Act all employers must treat men and women equally in employment, transfer and promotion and provide for equal opportunities such as access to additional occupational training and retraining. It is illegal to ask the applicant whether she is pregnant or plans to have children.

The Act expressly forbids job advertisements which state a preferred gender.

The Council of Equality (*Ligestillingsrådet*), set up under the 1988 Equality Act, aims to counter discrimination and promote equality between men and women. The council establishes studies and surveys to assess the extent of discrimination and generally acts to encourage the implementation of the Equal Opportunities Act and the Equal Pay Act. It does not take cases to court itself.

Equal pay

By the mid-1970s the principle of equal pay was adopted in trade union collective agreements and became consolidated in statute law. According to the 1989 Equal Pay Act (*Ligelønsloven*), employers who have taken on men and women at the same workplace must give them 'equal pay, including equal terms of pay for like work or work of equal value'.

The implementation of equal pay in Denmark has been conducted in several stages in the public as well as in the private sectors. The 1989 Act, which amended the 1976 Act, allows an individual employee to take legal action through the courts for equal pay. Claims based on non-fulfilment of a collective agreement which provides for equal pay must first be taken through mediation machinery. An employer who is covered by a collective agreement with a trade union has the duty to grant working conditions in accordance with the agreement, to non-unionized as well as to unionized employees. However, non-unionized employees cannot themselves bring a complaint regarding a breach of the rights laid down in the collective agreement. Provisions in an agreement which conflict with the principle of equal pay will be void. 'Pay' comprises everything that is laid out by an employer, directly and indirectly.

An employee who wishes to complain against breaches of these provisions must substantiate that there are differences in pay for the same work and show that they are due to sex discrimination.

Disabled and other minority groups

There are no statutes providing more favourable opportunities for the disabled in employment in the private sector.

Work environment

Safety

The Health and Safety at Work Act 1985 (*Arbejdsmiljøloven*) aims to promote a safe and sound working environment which is in consonance with current technical and social developments. It seeks to do so by enabling enterprises themselves to tackle health and safety issues through the labour market organizations and under the guidance and supervision of the Health and Safety Executive (*Arbejdstilsynet*).

The Act sets up a framework which is complemented by the Minister of Labour jointly with other labour market organizations.

The more specific regulations are issued as secondary legislation by the Minister of Labour. The Act contains provisions on:

- Instituting safety measures at the workplace.
- The rights and duties of the parties involved.
- The use of technical equipment.
- Substances and materials.
- Rest periods for employees.

Under the Act the employer is responsible for providing and maintaining working conditions that are safe and in accordance with the relevant laws and statutory instruments. The employer is also responsible for informing employees about any occupational hazards and providing them with the training and instruction necessary to carry out their work safely.

Enterprises are divided into three categories, each with a specific set of safety requirements. In companies with up to nine employees safety measures must be organized directly between employer and employees. In companies with at least ten employees one safety representative is elected by employees. A safety committee must be established where there are at least twenty employees. Departmental supervisors must oversee the safety measures within the company and guide and advise management on questions relating to health and safety.

Members of safety groups and committees are entitled to time off to acquire the knowledge necessary and to carry out their duties. Safety representatives enjoy the same protection as shop stewards in relation to dismissal and other disciplinary action. The employer is obliged to inform the safety representatives and shop stewards about any enforcement orders issued by the Health and Safety Executive and to co-operate with the safety groups and committees, an obligation also incumbent on employees.

Rest periods

The Health and Safety at Work Act provides for a continuous rest period of eleven hours per day and night. However, an exemption is made for shift work and agricultural work, where the rest period may be reduced to eight hours.

Liability

Violations of the Act may be punished by a fine or imprisonment. Some of the duties are on a strict liability basis: that is, an employer may be subject to punishment although the contravention was not intentional or negligent.

Individual termination of contract

For certain categories of employees, notably white-collar (salaried) employees, apprentices and vocational trainees, termination of contract of employment is subject to statutory regulation. For blue-collar workers, procedures are laid down by the various collective agreements. Additionally the conditions for the termination of the contract may also be laid down in the individual contract.

Grounds for termination

A fixed-term contract ends automatically when the time expires or the task is performed. Other contracts of employment normally come to an end upon suitable notice. The length of the period of notice depends on the regulations the employee is covered by and length of service (see below).

Grounds for dismissal relating to employee conduct range from failure to take up employment, unexcused leave, lack of competence, theft or violent behaviour. Grounds on which an employee may revoke the contract, and claim compensation, require a major breach of the contract by the employer, such as non-payment of wages or major changes in the conditions of employment without warning. The onus of proof is upon the party wishing to terminate the contract. In the case of individual dismissal there is no particular procedure distinguishing termination for reasons related to the person or conduct of the employee and termination on economic grounds.

Employees may give notice for any reason, whereas employers must adhere to article 4.3 of the General Agreement between the Danish Employers' Confederation and the Danish Confederation of Trade Unions, which protects the employee against arbitrary dismissals. Article 4 of the General Agreement provides for compensation for dismissals that are unwarranted by the circumstances of the employee or the undertaking. An assessment of whether a dismissal is deemed to be fair or unfair must be based on all the relevant and concrete circumstances related to the dismissal.

An employee default which is both substantial and gross may give grounds for summary dismissal without notice.

A dismissal will be deemed wrongful if it is prompted, for example, by an employee's membership of or activity on behalf of a trade union, an employee's religious or political affiliation, or the exercise of the employee's rights under the Maternity Leave Act or the Equal Pay Act.

Individual termination procedure

Notice to terminate the contract of employment will normally be in writing and contain details of the period of notice and any leave. Salaried employees are entitled to ask the reason for dismissal, as are blue-collar workers who have been in an undertaking for at least nine months. The party wishing to terminate the contract must ensure that notice has reached the other party by the required time. There are no consultation procedures for terminating the employment of an individual white-collar or blue-collar employee.

Period of notice

The length of the period of notice varies considerably, depending on the category of employee and length of service. There is a major difference in provisions for white-collar and for blue-collar employees.

For termination by the employer of the employment of a white-collar employee, the Salaried Employees Act provides that the employer must give at least:

- One month's notice, to expire at the end of the month, during the first six months of employment.
- Three months' notice, to expire at the end of a calendar month, after six months of employment.
- The period of notice above shall be increased by one month for every three years of service, subject to a maximum of six months.

A white-collar employee is required to give one month's notice, also to expire at the end of a calendar month.

Generally, in industry, under collective agreements blue-collar workers are entitled to two or three weeks' notice after one year's employment and four to seven weeks' notice after two to three years' employment. Long-service employees, with ten to twelve years' service, may be entitled to up to three or four months' notice. The employee may often terminate employment with only one week's notice.

Recourse to law, and settlements

Should the employee wish to contest a dismissal, the first step will be to enlist the support of their trade union, initially for enterprise-level mediation. If no agreement can be reached and the union agrees that the dismissal was unfair the union may either take the employer to the ordinary courts on behalf of the employee, or bring the case before the Tribunal of Arbitration on Dismissals set up under article 4 of the General Agreement. If the tribunal finds that a dismissal is unfair, and unwarranted by the circumstances of the worker or the undertaking, it may set the dismissal aside, unless there has been a breakdown in compatibility between the employer and the worker sufficient to preclude a continuation of the employment relationship. The tribunal may decide that the employer should pay compensation to the dismissed employee, based on service, which may not exceed payment equivalent to thirty-nine weeks. However, most complaints about wrongful dismissal appear before the ordinary courts, which generally fix lower damages for unfair dismissal. White-collar employees are entitled to severance pay under the law after long service: after twelve years' service the entitlement is one month's pay, two months' pay after fifteen years, and three months' pay after eighteen years in addition to pay during notice. However, employees have no claim to severance pay if they are entitled to a pension either from the employer or from the state.

Final formalities

At any time the salaried employee has the right to demand from the employer a written confirmation of:

- The duration of employment.
- The type of work principally performed.
- Remuneration received.
- The reason for notice.

Appendix

Ministry of Labour
International Section
19 Laksegade
DK-1063 Copenhagen K
tel. +45 33 92 59 00

The Labour Directorate
Finsenvej 78
DK-2000 Frederiksberg
tel. +45 31 10 60 11

Dansk Arbejdsgiverforening
(Danish Employers' Federation)
Vester Voldgade 113
Postboks 386
DK-1503 Copenhagen V
(publishes a guide to employment in Denmark in English)
tel. +45 33 93 40 00
fax. +45 33 12 29 76

Landsorganisation (the Danish confederation of trade unions)
Rosenørns Allé 12
DK-1634 Copenhagen V
tel. +45 31 35 35 41
fax. +45 31 37 37 41

Dansk Metalarbejderforbund
(National Metalworkers' Union)
Nyropsgade 38
Postboks 308
DK-1780 Copenhagen V
tel. +45 33 12 82 12

Council of Equality
Tordenskjoldsgade 27, 3
DK-1055 Copenhagen K
tel. +45 33 92 33 11

Health and Safety Executive
Svanevej 12
DK-2400 Copenhagen NV
tel. +45 31 81 55 55

3

France

Basic employer and employee rights and duties in France are set out in the constitution. Actual terms and conditions of employment are enshrined in a codified system of legislation, the *Code du Travail*. This sets out statutory provision on a large number of eventualities, although many of the rights it offers are qualified. For example, employees have the right to ask for training leave and employers have the right to refuse under certain circumstances. Working hours are also controlled by legislation, although the law enables employers to introduce a great deal of flexibility into working practices. The law does, however, regulate certain areas very strictly; recourse to temporary and fixed-term contracts is subject to firm statutory control.

Basic terms and conditions are often enhanced by sectoral collective agreements, covering most industries. These agreements may be extended by decree to cover all employees in a particular sector whether or not they have been party to the agreement.

Basic Rights

The French constitution states three basic principles with regard to employment:

- The right to work.
- Freedom of occupation.
- Professional equality between women and men.

The right to work

In order to realize the right to work laid down in the 1946 constitution, the 1958 constitution requires the state to play an active part in job creation, aid the training of the work force in order to meet the needs of the market, facilitate the placement of workers (notably through the state job placement agency, ANPE, Agence Nationale pour l'Emploi) and protect certain categories of workers.

Freedom of occupation

Although all French citizens have the right to choose their profession freely, there are certain types of employment which are prohibited. A person cannot legally work or offer a paid service independently without being first registered with the state employment agency, ANPE, and hence liable to corresponding tax and social charges. Civil servants and persons employed in public services may not engage in paid secondary employment; private-sector employees may do so only if it means that the total number of hours worked does not exceed the maximum number of hours allowed by law or collective agreement. The employment of foreigners, apart from EC citizens, is subject to qualitative and quantitative controls.

There are also some types of employment which are forbidden to certain categories of worker; for example, there are restrictions on the employment of children, notably in commercial institutions (law of 10 January 1970).

Professional equality between women and men

The principle of 'equal pay for equal work' is laid down in the 1946 constitution. A law on equality at the workplace enacted on 13 July 1983 brought French legislation into line with the 1976 EC directive on equal treatment. This subject is dealt with in greater detail below.

'Right of expression'

Under the law of 3 January 1986 all employees have a right to express opinions (*droit d'expression*) on a variety of topics such as the content, conditions and organization of work, proposed changes in working conditions and production or the introduction of new technology. This right mainly applies collectively but individual employees may express opinions and complaints themselves directly without having to consult works councils or employee representatives.

Freedom of association

Under Labour Code article L 412–1 employees are guaranteed freedom to join a trade union and to have trade union representation at the workplace (*droit syndicale dans l'entreprise*). This takes the form of trade union sections (*sections syndicales*) which have consultation and negotiation rights with the employer, or trade union representatives (*délégués syndicaux*), who may be appointed only in companies with fifty or more employees, unless otherwise stated in a collective agreement.

Union meetings may take place on company premises, but not directly at the place of work and not in company time. (Issues of collective rights and bargaining are dealt with in volumes 3 and 4 of the European Management Guides series).

Contracts of employment

The contract of employment is an agreement between employee and employer under which the employee offers to work under the direction of the employer in return for a wage based on time worked, piece rates or commission.

Employees have individual rights, defined partly by statute and partly by collective agreements. They include: the right to receive the contractually agreed salary; the right to be treated decently and in the same way as colleagues; and the right to the maintenance of the conditions set out in the contract of employment.

Employees are subject to a duty of fidelity to the employers and must not reveal confidential information or liaise with competitors. Employees are also bound to 'carry out, diligently and conscientiously' the tasks allotted to them and to take care of equipment and tools.

Employers must respect these rights and duties and are under an obligation to pay social security contributions, respect health and safety rules and provide the employee with a certificate of employment once the contract is terminated.

There are four main types of contract:

- Contracts of indefinite length (*contrats à durée indéterminée*).
- Fixed-term contracts (*contrats à durée déterminée*, CDD).
- Temporary contracts (*travail temporaire*).
- Work 'on call' (*contrat de travail intermittent*).

Contracts of indefinite length

Contracts of indefinite length are the most widely used type of contract and can be either written or verbal (Labour Code, article L 121–1). However, if they are written, they must be drawn up in French. Foreign employees may ask for a translation of the contract into their mother tongue, and case law has held that this is the valid text in the case of disagreement. Managers (*cadres*) generally receive written contracts. The areas which must be included in a written contract are often set out in a sectoral collective agreement. They would typically include a job description, the employee's classification and coefficient (see below), the length of the probationary period, if applicable, level of remuneration, working conditions and any other relevant employment conditions.

In France sectoral agreements allocate a number of points to each job title, depending on level of skill, complexity of tasks, etc. The greater the demands of the job in terms of these variables the higher the number of points. The total number of points associated with each level of job title is called the coefficient. Each point has a monetary value which forms the focus of bargaining during pay negotiations. The value of the point varies from industry to industry, just as the number of points allotted to each job varies. (The functioning of pay determination in France is dealt with in volume 4 of European Management Guides, on pay and benefits.)

Fixed-term and temporary contracts

The use of fixed-term and temporary contracts is strictly regulated. This follows concern about the steadily growing number of companies resorting to temporary staff instead of appointing permanent employees. The number of employees on these types of contract rose from 2·5 per cent of the total work force in 1977 to 7·1 per cent in April 1989, according to a Ministry of Labour report. An estimated 50 per cent of companies employ some staff on a temporary basis, and some companies are reported to employ 25–50 per cent of all staff on temporary or fixed-term contracts. This prompted the drafting of new legislation (Law No. 90-613), passed on 12 July 1990. The main provisions of the new legislation are set out below.

Fixed-term contracts must state the precise nature of the work and its duration, otherwise it is deemed to be a permanent contract. Such contracts should also contain the name and the qualifications of the employee to be replaced (if applicable), the finishing or renewal date, its minimum length (if no end date is given), the precise nature of the work, the collective agreement covering the company, the length of the probationary period (if applicable), the level of pay and any bonuses the employee is entitled to, and the name and address of the relevant pension and social security fund.

Both fixed-term and temporary contracts are renewable once only, and the total length of the contract may not exceed eighteen months. This can be extended, however, to 24 months under certain circumstances. These contracts must be written and presented to the employee within two days of starting work.

Employers must justify the use of temporary staff, and the legislation places strict controls on cases of recourse to it, limiting them to three broad categories as follows:

- Replacing an employee who is on sick leave or filling in a gap between the departure of an employee and the abolition of the post. Temporary employees must in no case be used to replace employees on strike.
- Coping with a temporary increase in work load, such as an exceptionally large order, the carrying out of a clearly defined and temporary task, or the carrying out of urgent work for health and safety reasons.
- Carrying out tasks of a temporary nature, such as seasonal work.

Employers who have dismissed employees for economic reasons may not make use of temporary or fixed-term contracts during the six months following the dismissal. Employees on fixed-term and temporary contracts may not be used to carry out dangerous work as listed by the Ministry of Labour.

Both temporary employees and those on fixed-term contracts are entitled to end-of-contract bonuses paid by the employer 'to compensate for the precarious nature of the employment'. From 1 November 1990 these bonuses amount to 5 per cent of gross salary for those on fixed-term contracts and 15 per cent for those on temporary contracts, reduced to 10 per cent if further employment can be found within three days. Employees on both temporary and fixed-term contracts are also entitled to a bonus in lieu of holiday pay, regardless of the duration of the contract. This bonus may not be less than the equivalent of a tenth of the total gross pay due to the employee, and is paid at the end of the contract.

Work 'on call'

Under legislation passed in 1986 (*ordonnance* 11.8.86), work on call 'may be concluded in order to carry out permanent work characterized by alternating periods of working and non-working'. This contract is of indefinite length and the employee is usually paid only for the periods worked. However, a collective or company agreement can provide for a monthly salary. For example, under the retailing and distribution collective agreement, companies may pay a monthly wage calculated on the agreed minimum working time, with overtime paid in the month it is worked. An employee on this type of contract has the same rights as a full-time employee. Non-worked periods count towards seniority. These contracts must be concluded in writing and include the qualifications of the employee, details of remuneration and a minimum *annual* working time. Actual hours worked may not exceed by more than a quarter this agreed minimum total. (That is, if 500 hours are agreed as a minimum, the actual hours worked may not exceed 625.)

In cases where work is liable to be unpredictable, provision for when employees may refuse to attend are laid down in collective agreements which are binding on non-signatory companies if extended by the Ministry of Labour.

Part-time contracts

Part-time workers are legally defined as those whose hours of work equal four-fifths or less of the statutory working week or the working week negotiated in an industry or company agreement. Works councils or workers' representatives must be consulted before part-time employees are engaged.

A part-time employee's contract must be in writing and state the hours to be worked. Any changes in working hours must be made known to the employee at least seven days before taking effect. Part-time workers have priority in consideration for full-time jobs, as and when they are created. An employee who works part-time has the same rights as a full-time employee.

If, during twelve consecutive weeks, overtime worked by a part-time employee exceeds the contractually agreed working time by more than two hours a week, the extra hours must be included in the contract if the employee so wishes. Total overtime worked must not exceed one-third of the weekly or monthly working time agreed in the contract.

Restrictive covenants

A contract may include a non-competition clause which prohibits the employee from engaging in any activities directly in competition with the current or former employer. The duration of the clause must be specified, as must the region of applicability and the precise kind of work defined. However, the clause must leave the employee free to 'carry on their profession normally', something which must often be decided by case law. These clauses are particularly common in the case of managerial employees and are sometimes provided for in collective agreements.

For example, the national agreement covering engineers and managers in iron and steel states that non-competition clauses must be set out in the initial contract of employment, cannot exceed one year and are renewable once only. In return the employee is paid a monthly sum equal to five-tenths of previous average monthly salary, including bonuses. If the employee is made redundant this is increased to six-tenths until the employee has found work. The employer can repeal the non-competition clause in a letter that must be sent no later than eight days following notification of the termination of the contract.

Employee inventions

By law any invention discovered by an employee during the course of their duties at the workplace is the property of the employer (law of 13 July 1978). If an employee has made an invention quite independently, without any help from the company, the invention belongs to the employee. Under the national agreement covering managers in the chemical sector, if an employee invents something, the employee's name must be mentioned in the patent application. However, this does not give the employee a right to co-ownership of the invention. If, within five years of the patent being granted, the invention is put to commercial use, the employee is entitled to a lump sum payment which is calculated according to the commercial value of the invention, the difficulty of putting it into operation and the part the employee played in its invention. The employee is entitled to this lump sum even if retired or no longer in the service of the company.

Individuals' workplace rights

Grievances

Grievances on the part of individual employees or apprentices against employers over their contract of employment or disputed dismissals are dealt with in the first instance by an industrial tribunal (*conseil de prud'hommes* — Labour Code, article L 514–2). Individual grievances between employees at the workplace are also dealt with by industrial tribunals (Labour Code, article L 511–1). Industrial tribunals cannot deal with collective grievances, although the same grievance presented separately by members of the work force will be treated as an individual grievance.

The first stage in the grievance procedure is the conciliation phase, in which both parties are heard in front of the conciliation service of the tribunal (*bureau de conciliation*), which will attempt to resolve the dispute. Both parties are advised of the hearing, which takes place in private, by letter. The hearing, which usually takes a day, can be attended by the parties in person or through a representative (e.g. a colleague, workers' representative or lawyer). If conciliation at this stage fails, the case is passed to the arbitration service of the tribunal (*bureau de jugement*), which will give a ruling. Both parties are officially notified of the hearing by letter, and the hearing is open to the public.

If still unresolved, grievances involving a total sum of more than Ff 15,000 (£1,515) would then pass to the court of appeal (*cour d'appel*) and then the high court of appeal (*cour de cassation*). Grievances involving a total sum lower than Ff 15,000 would pass straight to the high court of appeal.

Discipline

An employer may formally discipline an employee under Labour Code article L 122–40 which defines formal discipline (*sanctions disciplinaires*) as:

> all measures, excepting verbal warnings, taken by the employer following an act carried out by the employee which is perceived by the employer to be an act of misconduct [*agissement fautif*] . . . [which] will, immediately or otherwise, affect the presence of the employee in the company, their professional activities or remuneration.

Examples of acts of misconduct range from unauthorized absences from the place of work and consuming alcohol at the workplace to the divulgence of confidential information.

If an employer wishes to take disciplinary measures against an employee, the employee should be invited by registered letter to an interview to discuss the matter. In the interview the employer should explain the disciplinary measures being taken against the employee and the reasons for doing so. The employee may take along another member of staff to the interview if they wish. Disciplinary measures may not be taken less than a day or more than a month after the interview. Any disciplinary procedure must be commenced by the employer not more than two months following recognition of an act of misconduct.

Data protection

The type and amount of information which can be legally gained about an employee is limited under article 9 of the Civil Code, under which all persons have the right to respect for their privacy — an employer has no right to inquire into employees' private lives except when it is deemed relevant to the proper operation of the employer's business. Under law 78–17 the collection, storage and distribution of personal information by computer are regulated, and companies wishing to start up a database must notify the National Commission on Freedom of Computerized Information (*Commission Nationale de l'Informatique et des Libertés*). No persons, except next of kin and legal representatives, are allowed access to the personal identification numbers issued to all French citizens. Disclosure of information which would be obtainable only from a bank is a violation of article 378 of the Criminal Code. Public records, tax and property archives may only be consulted by the public seventy years after the date of the relevant entries.

Working time

Hours of work

The statutory working week was fixed at thirty-nine hours in 1982: the maximum working week over twelve consecutive weeks is forty-six hours, and the absolute maximum forty-eight hours in any one week. The statutory working week is defined as that time actually worked and does not, therefore, include either changing or washing time, meal or coffee breaks. In practice a collective agreement will determine whether such time is paid or not: the metalworking agreement, for example, excludes changing and meal breaks from the calculation of working time but includes washing time.

Some industry collective agreements may provide for a shorter working week. In chemicals, for instance, the working week is thirty-eight hours for non-shift workers, thirty-seven hours forty-five minutes for semi-continuous shift workers and thirty-seven hours thirty minutes for continuous shift workers.

The law recognizes that certain occupations involve discontinuous activity with an inevitable degree of idle time during the working day and in these cases allows the working week to be extended without additional pay. This so-called equivalence provision (*régime d'equivalence*) applies, for example, in restaurants and cafés, hairdressing and hospitals.

Under the law of 21 June 1936 the working week may be extended permanently with the payment of overtime for work which can be undertaken only outside normal working hours, such as maintenance work. Companies can apply for temporary extensions of the working week following temporary increases in work load.

There have been recent moves at national level to reduce the working week in order to create jobs. The national employment plan of 1989 introduced tax credits of Ff 1,000 (£101) per employee per hourly reduction for companies which reduce the working week by at least three hours.

The statutory maximum working day is ten hours, which can be raised, normally to twelve hours, during temporary increases in work load, on permission from the labour inspectorate and following consultation with employee and union representatives.

Working hours flexibility

Laws introduced in 1986 and 1987 permit a number of flexible working practices. Weekly hours worked in excess of the statutory working week (up to a limit of forty-four hours) can be offset against weeks when fewer than thirty-nine hours are worked, without employees being entitled to overtime payments or time off in compensation, or without the hours being counted towards the annual overtime quotas allocated to each industry (see below). Firms using continuous processes can also organize work in six to twelve-week cycles without payment of overtime, provided the working week averaged out over the period is no greater than thirty-nine hours. Companies may negotiate their own flexible working agreement as long as there is a collective agreement at either sectoral or company level.

A multi-industry framework agreement on flexible working, signed in March 1989 by unions and employers, contains guidelines on how to monitor and control flexible working time and recommends that the situation be regularly assessed at sectoral and company level.

The 1990 agreement on flexible working at the Banque de France gives an illustration of how this scope can and has been implemented in practice. It provides for flexible working hours between 07.30 and 19.00, with a minimum lunch break of forty-five minutes: employees can accumulate a maximum of two and a half hours a day or four and a half hours a week, over the agreed working week, up to a limit of ten hours, if there is sufficient volume of work to justify it. Similarly, employees may work shorter hours, accumulating debits of up to two and a half hours a day or four and a half hours a week, up to a maximum of seven and a half hours. Any hours' debits must be compensated for in the following month.

Women's night work

Under the Labour Code, article L 213–1, women are not allowed to carry out night work, which is defined as work undertaken during the period between 22.00 and 05.00, with certain exemptions, such as nurses. However, a special dispensation may be made in some cases; for example, after appropriate consultations, and with permission from the labour inspectorate, the consecutive seven-hour period counted as night work may be defined as any time between

22.00 and 07.00.

The 1987 law on flexible working hours allows exemptions to the ban on women's night work when 'because of particularly serious circumstances, it is required in the national interest', on condition that: an extended sector-level agreement provides for the possibility of such an extension to firms operating continuous shift systems; or a company or plant-level agreement sanctions the use of such an exception.

This legislation also lifts certain other restrictions on women's work, such as: the requirement to provide a one-hour break for a ten-hour period of working; the ban on women undertaking shift work; and the ban on women being employed in workshops or on sites during public holidays.

Employers who contravene restrictions on women's night work may be liable to fines of Ff 2,500 to Ff 10,000 (£252–£1,010).

Overtime

Overtime is counted from the fortieth weekly hour inclusive, up to an absolute maximum of nine hours a week or seven hours a week over a consecutive twelve-week period. However, employers may agree a number of hours of annual overtime per individual employee without needing the permission of the labour inspectorate. The national provision for annual overtime is 130 hours (national agreement of 17 July 1981) but sectoral collective agreements range from seventy hours in insurance to 188 hours in coal importing.

Employees are paid at time-and-a-quarter for the first eight hours' overtime and time-and-a-half from then on. Overtime which exceeds these negotiated limits requires prior permission from both the labour inspectorate and workplace representatives and must be compensated by thirty minutes' time off for each hour worked.

Time off

Annual and public holidays

Employees are entitled to a minimum five weeks' statutory annual paid holiday. There are eleven public holidays: New Year's Day, Easter Monday, May Day, Ascension Day, Victory Day (8 May),

Whit Monday, Bastille Day (14 July), Assumption, All Saints' Day, Armistice Day (11 November), and Christmas Day.

Sick leave

Employees with at least three years' service are entitled to thirty days' sick leave at 50 per cent of covered earnings from the fourth day of illness. This is increased to two-thirds of covered earnings after thirty days for employees with three or more dependent children. Most collective agreements improve upon this, providing for full pay for an initial period, typically one and a half months, followed by a further period at reduced earnings.

Maternity leave

Maternity leave normally covers a period of six weeks before the expected date of confinement and ten weeks after the actual date of the birth (Labour Code, article L 122-26) for women who have paid at least ten months' insurance contributions. Maternity leave may be extended to a maximum of ten weeks ante-natal and twenty-two weeks post-natal, depending on the number of children a woman already has and whether it is a multiple birth. Eighty four per cent of salary is paid by social security during maternity leave, with some sectoral employers (in chemicals and the wholesale trade, for example) making up the full salary. Adoption leave is ten weeks, beginning from the day the child joins the household.

Family events

Practically all employees are guaranteed the following paid time off (Labour Code, article L 226-1):

- Four days for the employee's own marriage.
- One day for the marriage of a child.
- Two days for the death of a spouse or child.
- One day for the death of a parent or sibling.
- Three days' paternal leave in the event of the birth or adoption of a child.
- Three days for pre-selection interview for military service.

Collective agreements, such as the one covering banking, may improve on these statutory provisions, providing for: ten days for marriage; two days for the marriage of a child; five days for the death of a spouse; three days for the death of a parent or a child.

Training leave

All employees have a right to individual job-related training, subject to certain conditions. Employees must normally have two years' seniority in the sector and six months' seniority in the company to apply for training leave (*congé–formation*). This leave must not exceed one year full-time or 1,200 hours if taken in stages or on a part-time basis. Employers may refuse applications for training leave, depending on the number of employees on leave at the time. The employee receives 80 per cent of normal salary during the period of training, which the employer can recoup from a joint fund.

Employees under 26 years of age with no previous training and three months' seniority in the company may apply for paid vocational training leave (*formation professionnelle*) of up to 200 hours a year.

Employees may apply for up to one year's unpaid leave (*congé d'enseignement ou de recherche*) if they have at least two years' seniority in a company. Employers can refuse the request if the absence amounts to 1 per cent of total hours worked in companies with less than 200 employees or if more than 1 per cent of the work force would be absent in companies of over 200 employees.

Under the 9 July 1970 agreement on training, employees made redundant may take a minimum of two months' paid leave for training if they have two years' seniority.

The social partners are to meet during 1991 to negotiate an agreement on training, providing for regular meetings to discuss the structure and financing of training. It is expected that a new law on training will be drafted in 1991 to replace the existing law of 1971.

Employee representatives

Employees with employee, union or health and safety representative duties are entitled to paid time off ranging from two to twenty hours per month, depending on the nature of the duties and the size of the company.

Unpaid leave of absence

Employees may suspend their contract of employment for six to eleven months without pay to go on sabbatical leave (law of 3 January 1984) or for one or two years in order to set up a business as long as the employee has thirty-six months' length of service in the company and a total of six years in employment.

Retirement

State pensions

Normal retirement age for women and men has been 60 years since 1 April 1983. The state pension scheme for non-managers, ARRCO, guarantees 50 per cent of former annual earnings, based on an employee's ten best-remunerated years as long as they have paid social security contributions for at least thirty-seven and a half years. For employees with fewer than thirty-seven and a half years' contributions the pension is reduced on a pro-rata basis to a minimum of 25 per cent of former earnings after thirty-two and a half years' contributions. Whilst employees may retire at the age of 60 if they wish, the option is voluntary and they may continue working, if desired, in order to make up their full state entitlement.

The compulsory supplementary pension scheme (*retraites complémentaires*) guarantees employees an average of 20 per cent of former earnings (also based on the best ten years), bringing income from the state and supplementary schemes up to a total of 70 per cent of former earnings. This supplementary pension scheme, financed by the unemployment fund UNEDIC, is set to run until 1 January 1993.

The pension scheme for managers, AGIRC, covers some two million supervisory and managerial employees. It operates on the basis of 'pension points' which accrue according to employee contributions. Contribution rates vary according to salary.

Early retirement

Although the government scheme of 'solidarity contracts' for employees retiring after the age of 55 ended on 31 December 1983, solidarity contracts can still be signed on phased early retirement.

Employees aged less than 60 whose job is turned into two or more part-time jobs will be reimbursed by the government for 30 per cent of average gross pay lost when they voluntarily accept the introduction of job-sharing. The reimbursement continues until the employee draws the state retirement pension.

Firms with special economic difficulties may also, with state agreement, retire people over the age of 56 years and 2 months with a guaranteed income of 65 per cent of salary until the age of 60.

Equal treatment

Sex equality

Equality of pay and treatment between men and women is a statutory obligation, with contravention punishable by fine or imprisonment. On 13 July 1983 a new law on equality at the workplace was enacted, referred to as the 'Roudy law', after Yvette Roudy, then Minister of Women's Rights. This law brings French legislation into line with the 1976 EEC directive on equal treatment and also provides a framework for employers and unions to promote equality in all areas of working life. The law defines 'work of equal value' as jobs that are broadly comparable in terms of professional skill and experience required, responsibility and mental or physical effort.

The legislation also contains the provision that all companies with fifty or more employees must draw up an annual report on the comparative situation of women and men in the company with regard to recruitment, training, promotion, job qualifications, job grades, employment conditions and pay. Details must also be given concerning steps taken during the previous year to ensure equality, and measures planned for the coming year. This report is submitted to the works council or, where none exists, to the employee representatives of the company and is sent to the local labour inspectorate together with the opinion given by the works council.

In 1987 the Ministry of Social Affairs made financial assistance available on an experimental basis to small firms of under 200 employees in order for them to offer women individual training contracts to enable them to undertake jobs previously held by men. However, the number of companies adopting the scheme did not increase as much as expected over 1988 and 1989, the number of companies taking advantage of the scheme only rising to twenty-four in 1989 from nineteen in 1988. The majority of companies involved are reported as putting forward in-house training schemes already in existence, instead of devising new training schemes, in order to benefit from the subsidies.

At industry level, a multi-industry agreement on sex equality (*accord interprofessionnel sur l'égalité professionelle*) was signed by the main employers' organization, CNPF (Confédération Nationale du Patronat Français), and union confederations in November 1989. The agreement reaffirms the principle of 'equal pay for equal work' and concentrates on the promotion of women's employment and equality in the workplace. The agreement's main provisions include:

- Bringing collective agreements into line with the EC directive (76/207) on equal treatment of men and women.
- A commitment to initial and continuous professional training as a means of ensuring equal opportunities, especially after a period of parental leave.
- Better information concerning the situation of women in the labour market.
- A commitment to making it easier for employees of both sexes to combine work and family life, particularly through provisions to ensure the care of a sick child.

However, according to a Labour Ministry survey, the agreement has had a limited effect on the modification of collective agreements. The survey looked at sixteen collective agreements which had been renegotiated in 1989 and found that, of the five which mentioned the principle of sex equality, all had already included the principle in the former agreement. Eleven agreements contained provisions for caring for a sick child. Although only four of them extended the provision to fathers, no agreement expressly stated that this benefit was reserved for mothers. Eleven agreements contained provisions for leave to bring up a child (*congé pour élever un enfant*) for both mother and father, five of these revising clauses which had formerly reserved this right for mothers only. Many referred to the Labour Code articles L 122-28-1 and L 122-28-4 on parental leave and the possibility of returning to work on a part-time basis.

Disabled workers (*travailleurs handicappés et mutilés de guerre*)

By law, disabled employees enjoy a certain amount of protection with regard to terms and conditions of employment. Under legislation passed in 1987 companies with at least twenty employees must employ a number of disabled staff equivalent to not less than 6 per cent of the total work force as from 1 January 1991, or pay a sum of Ff 15,000 (£1,515) to the national association for the funding of the employment of the disabled (AGEFIPH). Companies may apply for various state subsidies available for employing and training the disabled.

At sectoral level an agreement on the employment of disabled workers was signed in the insurance sector in November 1989 by employers and unions. The agreement provides for a fund to encourage companies to take on disabled people for a period of training (*stage*) and a second fund to improve disabled access to the workplace and the availability of special transport. In the case of dismissal for economic reasons (*licenciement économique*) employers are urged to take into account an employee's disability as much as possible and try not to reduce the number of disabled staff employed if possible.

At industry level the national electricity and gas industries' (EDF–GDF) three-year agreement was signed in November 1989. It pledges the annual recruitment of a number of disabled workers equivalent to at least 1·5 per cent of the overall level of recruitment, around seventy persons a year. A total of 50,000 hours a year will be made available for the initial training of disabled workers in 1991, up from 30,000 in 1990.

Ethnic minorities

There is statutory provision for equality of treatment between ethnic minorities and French nationals with regard to workplace rights, working conditions and social protection. Under the decree of 17 August 1982, if collective agreements are extended across a sector, they must include the principle of 'equality of treatment between French nationals and foreign workers, notably in matters relating to employment'. Foreign workers are assured equality in matters relating to pay (and the right to send money home) and cannot be dismissed on grounds of ethnic origin, race or religious convictions (Labour Code, article L 122-45).

Work environment

Health and safety

By law (Labour Code, article 236), all companies employing at least fifty staff must establish a health and safety committee (*comité d'hygiène, de sécurité et des conditions de travail*). Members of the committee are entitled to time off, ranging from two to twenty hours per month, depending on the size of the company. Companies employing fewer than fifty staff are exempt, although the labour inspectorate can insist in some circumstances, depending on the nature of the work or the workplace. In the absence of a health and safety committee, workers' representatives (*délégués du personnel*) will cover duties relating to health and safety matters.

Employers are under obligation to keep the work place clean and hygienic (Labour Code, article L 232-1). Places of work must be adequately ventilated, heated and lit.

Under the Labour Code, article 233-1, employers have an obligation to protect employees against falls, collisions and suffocation with regard to the use of stairs, scaffolding, lifts and walkways, in the workplace in general or in the case of any construction work being undertaken in the workplace. Any potentially dangerous installations such as wells, vats or water tanks must be properly covered, and machines must be properly installed and maintained. Under articles 231-6 and 231-7 all dangerous substances must be clearly labelled as such and carry proper instructions for use.

By law (Labour Code, articles L 231-8, 8-1 and 8-2) all employees have the right to stop work if they feel that their health or safety is at risk in a certain situation (but not to turn their machines off) and tell their employer or representative. The employer cannot request the employee to recommence work as long as the danger persists. The works council may bring in experts if the employer does not agree that there is a serious danger.

Prevention of fire

There is a good deal of legislation (R 233-17-41) concerning fire exits, the maintenance of heating and lighting installations in order to prevent the outbreak of a fire, and dealing with fires. In buildings that house at least fifty people a fire information point (*consigne d'incendie*) must be established, and notified to the labour inspectorate.

Noise

Under the decree of 21 April 1988 (88-405), employers must measure and record noise levels at the work place every three years. The labour inspectorate has the right to inspect the records.

Food and drink

By law (L 232-2, 3), employers are under an obligation to provide employees with drinking water facilities and, for certain types of work entailing physical exertion, free non-alcoholic drinks. The employer is obliged to provide a canteen if the number of workers desiring to eat on the premises is twenty-five or more. If the number is less, the employer must provide a clean, hygienic place for employees to eat, away from the actual place of work.

Individual termination of employment

There are two main grounds for termination of contract:

- For reasons related to the employee.
- For economic reasons.

Dismissal for reasons related to the employee

Dismissal for reasons related to the employee can follow from one of three things:

- Flagrant misconduct (*faute lourde*), following unacceptable behaviour on the part of an employee, such as fraud, industrial espionage or theft.
- Gross misconduct (*faute grave*), following behaviour which precludes the continued presence of the employee in the company, such as insubordination or harming the security or interests of the company.
- All other cases, which include personal reasons such as personality clashes, continual aggressiveness, or professional reasons such as incompetence or poor performance.

The first two reasons are grounds for summary dismissal, but for the third category notice with severance and holiday pay are required. Employees with at least two years' service, on permanent contracts, who have not been summarily dismissed are entitled to a legal minimum severance payment of one-tenth of a month's salary for each year of service plus an additional payment of one-fifteenth of a month's salary for each year of service over ten years. This is a lump sum and is exempt from tax and social security contributions.

Dismissal on economic grounds

Dismissal on economic grounds is now legally defined, under legislation passed on 2 August 1989, as 'dismissal by an employer for one or more reasons not related to the person or the workers concerned, resulting from the disappearance or alteration of (their) employment or a substantial modification of the employment contract as a result of economic difficulties or technological change'.

If the employee requests it in writing the employer is obliged to give real and well founded reasons (*causes réelles et sérieuses*) for all dismissals, whether economic or non-economic (Labour Code, article L 122-14-2).

Period of notice

In the case of an employee resigning, no specific period of notice must be given, although it may be specified by collective agreement or custom and practice within the industry. If an employee is dismissed, they are entitled to notice periods if they have six months' service and are not summarily dismissed. Statutory notice periods are one month for employees with six months' to two years' service and two months for employees with at least two years' service.

However, some collective agreements provide for longer periods of notice: for example, the national managerial and engineers' agreement for metalworking, which provides for three months' notice for all but the lowest grades of staff after a probationary period of at least three months.

Termination procedure

Termination for non-economic reasons. Termination procedures in France are very rigid. Fines can be imposed for failure to keep to the formalities. The following procedure is applicable to employees in all establishments, regardless of size, the employee's length of service or whether the dismissal concerns a *faute lourde*, a *faute grave* or some other grounds.

- A registered letter with a slip to acknowledge its receipt must be sent by the employer to the employee requesting attendance at an interview (*entretien préalable*). Alternatively, employees may be informed by means of a letter put into their hands for which they sign to acknowledge receipt. The letter must specify the purpose, date, time and place of the interview, as well as the fact that employees may be accompanied by a person of their choice who is employed by the company.
- An interview is held at which the employer explains to the employee the grounds for the dismissal; the employee explains his or her side of the story.
- If an employee is dismissed on personal grounds, a registered letter must be sent to the employee, confirming the dismissal and the reasons for it, no earlier than one full day after the interview. If an employee is being dismissed for disciplinary reasons the letter must be sent within the month following the interview. Under the August 1989 legislation the employer must state in this letter the full reasons for the dismissal. Previously this was the case only with dismissals for economic reasons, although employers were obliged to state the full reasons if requested by the employee (see above).

Termination for economic reasons. Under the Labour Code, article 321-2, there is no employer obligation to consult the works council or employee representatives in the case of individual dismissals for economic reasons (the obligation starts with the dismissal of two or more employees).

- As with dismissals for non-economic reasons, the employee is requested to attend an interview during which the employee is offered a retraining contract (*convention de conversion*). An employee has twenty-one days in which to decide whether or not to take up the offer. Retraining contracts are offered individually to employees whose jobs have been eliminated, to enable them to learn new skills and gain further qualifications to facilitate future job placement. The scheme is jointly managed and financed by the state (through the national employment fund), the unemployment insurance organization and companies.

- A registered letter confirming the dismissal may be sent to the employee during the twenty-one days following the interview, but no earlier than seven days following the date of the interview. If the employee makes no reply to the offer of a retraining contract during those twenty-one days, it is assumed that the offer has been rejected.
- Employees dismissed for economic reasons have priority of reinstatement for a period of one year dating from the termination of their contract provided that they make the wish to be reinstated known to the employer within four months.

Unfair dismissal

If an employee feels that they have been unfairly dismissed, they must appeal through a labour tribunal (*le tribunal*). An employer may be liable for damages if it is found that the employee has been unjustifiably dismissed (*pour une cause non réele et sérieuse*) or if the employer has not followed the exact procedures in effecting the dismissal (Labour Code, article L 122-14-4). The penalties are as follows:

- If an employee with at least two years' service in a company with at least eleven employees is unjustifiably dismissed, the tribunal may propose — but not enforce — the reinstatement of the employee with all his or her rights intact. However, if either party refuses to accept reinstatement, then the employer is obliged to pay the employee a sum in compensation not lower than the previous six months' earnings. The employer must also reimburse the state for any unemployment benefit claimed by the employee between the date of dismissal and the date of the tribunal's judgement.
- If the employer is found to have failed to follow the correct dismissal formalities the employer must pay the employee, at most, one month's wages in compensation, but is not required to reimburse unemployment benefit. In this case the judge may not propose reinstatement.

These provisions do not apply to employees with less than two years' service or who work in companies with fewer than eleven employees. Such employees may, however, take civil action against their employer where appropriate, and it is likely that the court would award compensation equivalent to six months' earnings if unfair dismissal were upheld.

Ending the employment relationship

Employee rights and obligations

Employees who resign are required to abide by the periods of notice fixed by collective agreement or by custom and practice for a particular job or profession (see above). An employee who resigns is not bound by the periods of notice which an employer would have to recognize by law were the employee dismissed or made redundant.

Any employees who suddenly break the employment contract without following the required period of notice risk losing to the employer a sum in lieu of notice (*une indemnité de brusque rupture*) equal to the earnings they would have received had the period of notice been served out. An employee who breaks the contract with wilful intent may be liable for the payment of damages to the employer.

Employer rights and obligations

If an employee resigns the employer must allow the employee to serve out their period of notice doing exactly the same kind of work as before. Any attempt by the employer to alter working conditions unilaterally — or not to allow the employee to serve the period of notice — gives the employee the right to claim payment in lieu of notice (*indemnité compensatrice*).

Once an employment contract has expired for any reason, the employer must:

- Give the former employee their employment certificate (*certificat du travail*), which contains the dates relating to the duration of the employment contract, a description of the positions occupied by the employee in the company and the dates during which they were occupied (Labour Code, article L 122-16).
- Pay the employee all outstanding sums owing, such as salary, bonuses, any payment in lieu of notice, and so on. The employer

should ask for a receipt for all payments due. The employee has two months from signing the receipt in which to question it.

Appendix

Ministry of Labour
127 rue de Grenelle
75700 Paris
tel. +33 1 40 56 60 00
fax. +33 1 40 56 67 44

Agence Nationale pour l'Emploi (ANPE)
23 rue Félix
75009 Paris
tel. +33 1 49 31 74 00

Confédération Générale des Cadres
(national executives' union)
30 rue de Gramont
75002 Paris
tel. +33 1 42 61 81 76

Direction des relations du travail et la protection contre les risques du travail (Health and Safety Directorate)
127 rue du Grenelle
75700 Paris
tel. +33 1 40 56 60 00

Conseil National du Patronat Français
(national employers' organization)
31 rue Pierre I de Serbie
75016 Paris
tel. +33 1 40 69 44 44

British Chamber of Commerce
8 rue Cimerosa
75118 Paris
tel. +33 1 45 05 13 08
fax. +33 1 45 53 02 87

Commission Nationale de l'Informatique et des Libertés
21 rue de St Guillaume
75007 Paris
tel. +33 1 45 48 39 39

4

Germany

The regulation of terms and conditions of employment in Germany is derived from a number of constitutional principles, many of which concern collective rights, and many uncodified individual pieces of legislation. Some of these, such as the Civil Code, the Industrial Code and the Commercial Code, date back to the industrial and commercial legislation of the late nineteenth century, although all have been subject to numerous amendments since. Most legislation in the employment field now exists and is introduced at federal level, although in theory the constituent *Länder* do possess competence in this field. Apart from educational leave, little *Land* legislation affects employment relations.

This chapter deals with the laws, collective agreements and practices applicable in the former Federal Republic of Germany. Most civil law, including labour law, also applies in the five new *Länder*, the former German Democratic Republic. Some transitional arrangements will remain in force in the new *Länder* during 1991, notably on short-time and early retirement, and other former GDR provisions will continue until all-German law is revised in some areas.

Basic rights in employment are those granted in the Basic Law, the founding constitution of the Federal Republic of Germany, promulgated in 1949. The bases of the law of contract and damages, which have also been cited in sex discrimination cases, are found in the Civil Code. Building on these principles, the courts have elaborated a body of case law which has been decisive in shaping the law governing collective bargaining and industrial action, and privacy and data protection. The Basic Law establishes basic rights of liberty and the inviolability of the individual, and freedom of occupation. Article 9 grants freedom of association, and expressly the right 'to form associations to safeguard and improve working and economic conditions'.

Contracts of employment

German law sees the contract of employment (*Arbeitsvertrag*) as one form of the wider notion of a contract of service (*Dienstvertrag*) and differentiates it from a contract for services (*Werkvertrag*) through the fact that the latter must entail the supply of finished goods or a service in contrast to the performance of a specified activity under the supervision of and for an employer.

Types of employee

The status of 'employee' confers rights and duties *vis-à-vis* the employer, and brings the individual within the scope of labour law and the jurisdiction of the labour courts. The number of employees in an establishment determines co-determination rights, both at establishment and at board level, as well as rights under the 1969 Protection against Dismissals Act. There are two basic categories of employee: manual, or blue-collar (*Arbeiter*), and white-collar (*Angestellte*). Within the white-collar group 'executives' (*leitende Angestellte*) are subject to separate provisions as regards termination, and do not count as employees for the purposes of the Works Constitution Act 1972. Executives are employees who may:

- Hire or dismiss employees on their own authority, or who have general authority to conduct business on behalf of their company, or
- Regularly carry out tasks of significance for the existence or development of the enterprise or establishment, requiring particular experience and knowledge, largely free of direct instruction.

In the event of doubt, an executive is anyone:

- Previously assigned to this group in a work place election.
- Employed at a level of management at which most other employees are classified as 'executives'.
- In receipt of a salary which is the norm for executives in that enterprise or three times the social security calculations base (in 1991 DM 120,960 or £41,000).

Apart from differences in periods of notice and in sick pay, most major legal differences between these categories have been removed, or are intended to be removed (see periods of notice, below). However, the groups remain separate for the purpose of works councils election.

Form of contract

The contract of employment does not require a particular form except for certain categories of employee, principally trainees: it can be oral, in writing, explicit or implied, and does not have to specify every detail. In the event of a disagreement the burden of proof lies with the party seeking to obtain their rights under it. The presentation of working papers by the employee and deduction of tax and social insurance contributions by the employer are regarded as denoting that a contract of employment has been concluded.

Collective agreements (usually at industry level) or works agreements (at establishment level) may require that a contract of employment be in writing, or that employees should receive written notification of engagement and basic terms.

A contract may be null and void where one of the parties has 'erred as to a characteristic of the person or of a matter which is essential to the transaction' (Civil Code, section 119); where either the employer or the employee has engaged in malicious deception or threats; if the contract breaks a law, violates 'good morals' or is for an illegal purpose. Contracts with minors need the permission of their legal guardian. The 1869 Industrial Code gives employer and employee freedom to settle the terms of a contract, subject to statute law. Since collective agreements (*Tarifverträge*) and works agreements (*Betriebsvereinbarungen*) under the Works Constitution Act have the force of law, individual agreements must conform with them where applicable. Individual terms may diverge if they are favourable to the employee.

A written contract of employment, standard forms for which are widely available in bookshops, would specify:

- Department and grade (if appropriate).
- Probationary period (see below).
- Agreement regulating transfer of the employee within the firm.
- Remuneration.
- Working hours, leave and time off (possibly by reference to any applicable collective agreement).

- Employee obligation in the event of prevention of performance.
- Period of notice (may be set by collective agreement).

Types of contract of employment

The most common form of contract of employment is for an *indefinite* term. The parties to the contract may agree a probationary period. This can take two forms. In the first a probationary period precedes an indefinite contract of employment. There is no statutory provision on the length of a probationary period, and in any case, an employment contract can be terminated during the first six months of employment without dismissal protection legislation being invoked. Collective agreements set probationary periods, which may vary, depending on the category or level of employee. For example, the framework agreement for the metalworking industry in the North Württemberg/North Baden region allows an eight-week probationary period for blue-collar and a three-month probationary period for white-collar employees, extendable to six months in the case of certain white-collar grades with the consent of the works council. A probationary period can be terminated, subject to the minimum statutory periods of notice: that is, two weeks for a blue-collar employee and one month for white-collar employees. A probationary period can also take the form of a fixed-term contract.

Fixed-term contracts

Under a *fixed-term* contract employees are not subject to ordinary dismissal: they can usually be dismissed only summarily. On the other hand, employees have no protection against the termination of the agreement when the contract runs out. In general a fixed-term agreement can be concluded only if there is 'material reason'. This would include: probationary employment, filling a temporary vacancy, a specific task limited in time, or at the employee's request. The length of a fixed-term contract is subject to the same considerations which determine the existence of a 'material reason'. In the case of scientific or technical staff workers on long-term projects, up to four years might be agreed. Fixed-term contracts may be renewed repeatedly only for 'telling material reasons'.

Under the Employment Promotion Act 1985, which will remain in force until 1995, a fixed-term contract can be agreed *once* for up to eighteen months without 'material reason' in the case of a new employee or after training where no indefinite post is available. This eighteen-month period can be extended to two years if the employer has begun business less than six months previously or if the employer has ten or fewer employees, excluding trainees. Case law has ruled that a series of fixed-term contracts can be agreed under the Act — that is, without a 'material reason' — if four months elapse between the end of one contract and the beginning of the next and there is 'no close material relation' between successive periods of employment.

Part-time employment

Part-time work is any period of work less than that worked by full-time employees in an establishment. Some collective agreements specify minimum periods for part-time work, together with preferential treatment when applying for full-time posts. Part-timers who work less than fifteen hours per week and earn a monthly income of less than DM 480 (£163) are not subject to sickness and pensions insurance contributions and benefits. Part-timers who work less than nineteen hours are excluded from unemployment insurance and benefits. Employees who work less than ten hours per week or forty-five hours per month do not count for the purposes of the minimum number of employees required to bring an establishment within the scope of the Protection against Dismissals Act. Employers are liable for all the social security contributions of employees earning less than DM 610 (£207) per month.

Part-timers must not be treated differently from full-time employees without a material reason. Nor must differential treatment constitute indirect sex discrimination. Recent judgements of the European Court of Justice dealing with pensions and sick pay have now been confirmed by the German courts.

Work 'on call'

Work 'on call', also known as 'capacity-orientated variable working time' (KAPOVAZ), requires employees to work when requested to do so by the employer. Such a contract must specify the length of the working week, and if no period is given, then a minimum period of ten hours will apply, for which the employer must pay the employee. Legal opinion holds that the weekly working time may be averaged over a year, although there is still uncertainty about how the law should be interpreted. Employers must inform employees at least four days before a period of work and employ them for a minimum of three successive hours on any workday. Work 'on call' requires the consent of the works council (with right of appeal to conciliation).

Job-sharing

Under a job-sharing contract, each of the partners is required to specify the period of work for which they are responsible. It is unlawful for a contract to require that the post be constantly occupied — that is, that one partner should deputize for the other in the case of absence. However, a general agreement to deputize in the event of 'serious operating exigencies' can be made.

Rights and duties under the employment contract

Section 611 of the Civil Code obliges the employee to perform 'the promised services' agreed in the contract of employment, and the employer to provide 'the agreed remuneration'. Where the contract is not specific, the employee's tasks can be determined by the employer exercising the right of direction, subject to law and reasonableness. The courts have upheld an employee's right to refuse certain types of work if it conflicts with their conscience, and employers have been required to make other work available.

An employer may not transfer an employee to lower-paid work unless it is allowed for in the contract: should an employer wish to change an employee's terms of employment, the employer must go through the procedure for conditional termination of employment to change a contract's terms. In establishments with twenty or more employees transfers of employees must be notified to the works council.

In general, employees are subject to a duty of fidelity to their employer, and employers have a duty of care towards employees. This requires employers to ensure, for example, that workrooms, equipment and the organization of work do not constitute a hazard to employees' life and health. Employers are also responsible for ensuring the safety of employees' property, and for providing secure areas where belongings may be kept during working hours.

This obligation also extends to the employee's rights as an individual (for example, employers may not bug employees' telephones, or secretly observe them with videos, and must protect information about them which should be confidential). Equipment intended to monitor employee performance cannot be introduced without the agreement of the works council.

An employee may take on other employment provided that this is not expressly ruled out in the contract of employment, does not involve unfair competition with the employer, does not prejudice employees' ability to work for their first employer, and does not entail exceeding the statutory maximum working week, currently forty-eight hours. Employees are forbidden to divulge commercial secrets for the purposes of competition, for their own gain or with intent to damage the interest of their employer. There is also a broad requirement of confidentiality covering trade and business matters such as pricing, the company's credit rating, technical know-how and financial information. However, once employment is terminated, employees may use any knowledge and experience honestly acquired in the course of the employment where it is reasonable to do so in order to further their career.

Restrictive covenants

A restrictive covenant to cover the period after employment has been terminated is subject to a number of conditions. Former employees must be compensated for agreeing to a restraint, the sum to be not less than 50 per cent of their former remuneration. (This sum must be offset against any earnings from subsequent employment where current earnings and compensation for agreeing to the covenant exceed 110 per cent of former earnings or 125 per cent if the employee has moved residence.) A restrictive covenant must be in writing, last no longer than two years, and serve a 'legitimate interest' of the employer. A restrictive covenant is not binding if it represents an unfair obstruction to the employee's career advancement. The spatial extent of the restraint must also be agreed, and must be reasonable. If the employer has breached the contract of employment, leading to termination by the employee ('constructive dismissal'), the employee may release him- or herself from any restraint. A 'socially unwarranted dismissal' (see below) may also release an employee from the obligation.

Individuals' workplace rights

Many individual rights, such as privacy or freedom of conscience, are protected by the general principles outlined above. Particular statutes also grant other individual rights.

Under the Works Constitution Act 1972, which regulates the election and rights of works councils in establishments with five or more eligible employees, individual employees are entitled to treatment 'in accordance with the principles of law and equity', and in particular are protected against discrimination on grounds of race, creed, nationality, political or trade union activity or beliefs, sex or age.

Although much of the Works Constitution Act regulates collective rights, dealt with in volume 3 of the European Management Guides series, certain individual rights are also guaranteed. For example:

- Employers must inform employees of their tasks and responsibilities, and of any associated health hazards.
- Employees have a right to be heard by any manager with responsibility for their work, to state their case on any measure concerning them, and to make suggestions on the design of the work place and the organization of operations.
- Employees are entitled to an explanation of how their remuneration is calculated, and may also apply for an interview with management on their performance and career prospects.
- Employees have access to their personal files, and may append their own observations.
- There is a right of complaint in the case of employees who feel they have suffered discrimination or unfair treatment.

Employee grievances would normally be routed through the works council or the employee's immediate superior. The works council may take a grievance to the conciliation machinery if it fails to resolve the problem. Company personnel manuals may prescribe or detail grievance arrangements along these lines, but may in addition grant direct employee access to more senior management.

Working time

Hours of work have been a central issue in West Germany over the past decade. Ever since the late 1970s the trade unions have made cuts in working hours a bargaining priority. Despite initial differences over whether they should take the form of a shorter working week or of a reduced working life, bargaining has recently concentrated on cutting the working week. The first agreements providing for a thirty-five-hour week by 1995 were concluded in the metalworking and printing industries in 1990. There have also been developments in time-off provisions, notably in educational and parental leave.

Hours of work

The current law on working hours, the 'Working Time Ordinance', dates from 1938, and the Conservative–Liberal coalition government has stated that it intends to reform it during its present term of office. The law sets a maximum working week of forty-eight hours, based on a maximum of eight hours a day over six working days, with Saturday counting as a workday. Work on one day, or several days, may be shortened, and the remaining hours distributed over other days, provided no more than ten hours a day are worked. Subject to collective agreement, agreement at plant level, or where appropriate a ruling by the local labour inspectorate, this possibility allows for considerable theoretical flexibility in working schedules. For example, flexi-time is both permitted and quite common, although, unless otherwise agreed collectively, the prescribed average working week must be achieved within a three-week reference period.

In practice, collective agreements regulate the working hours and patterns of most employees. Most industries in West Germany now work between thirty-seven and thirty-nine hours a week. Nevertheless the direct effect of legislation is still important for some groups of employees (see below). Hours in the East are longer, with moves to a 40-hour week in most industries under negotiation in 1990/91.

Although the length of working time is set primarily by collective agreement, the specific configuration of working hours, shift patterns, and starting and finishing times are all subject to co-determination by the works council in an establishment, if one has been elected by the work force.

Sunday and weekend working is regulated by the Trade Code 1869. Apart from work in restaurants, places of entertainment, and in transport, industrial work on Sundays or public holidays is not permitted except for essential maintenance, to meet 'the daily needs of the population', for seasonal work, where the work by its nature does not permit an interruption, or where continuous operation is necessary to prevent a deterioration of raw materials or defective output on technical, but not economic, grounds.

The former GDR law on Sunday working, which allows fewer exemptions, will continue in force until 31 December 1992.

Shop workers. Although the length of the working week is set by collective agreement, shop opening hours are controlled by special legislation, the Shop Closing Act (*Ladenschlussgesetz*), most recently amended in 1989. Shops — except those on railway stations or at airports — must close on Sundays and public holidays, and on Mondays to Fridays between 18.30 and 07.00. Under the recent amendment, part of the government's effort to liberalize business hours, shops may remain open until 20.30 on Thursdays, but must not stay open any longer than the total number of hours available under the 07.00–18.30 opening rule. Shops must close at 14.00 on Saturdays, except on the first Saturday of the month, the so-called 'Long Saturday', when they can stay open until 16.00 between April and September, and until 18.00 on the four Saturdays before Christmas. This deregulation was fought by the retail and commerce trade union HBV, which succeeded in gaining a non-binding commitment from employers to normally close at 18.30, unless they are confronted by competition from retailers who were not party to collective bargaining. Although the longer Thursday opening has drawn more people into city centres, many retailers remain sceptical about whether the additional costs have been warranted by extra sales.

Young people under 18 may not work more than eight hours a day, or forty hours a week, and must have a 12-hour break between shifts. Daily working hours can be lengthened by half an hour, for example within a flexi-time system, where hours are shortened on other days.

Overtime

Although overtime is covered by the 1938 ordinance, in practice its limits are so broad that they do not provide a major constraint on overtime working. Overtime within the meaning of the law — termed *Mehrarbeit* — needs to be distinguished from overtime beyond the individual's usual hours of work — termed *Überstunden*. Where overtime within the meaning of the law is worked, it must be paid with a supplement of at least 25 per cent. Collective agreements often set lower limits, specify premium rates and provide for time off in lieu. In the metalworking industry, for example, up to ten hours' overtime a week or twenty a month can be worked by agreement with works councils to meet 'urgent operating requirements'. Works councils enjoy an enforceable right of co-determination over the introduction of overtime, which means that if they should refuse a request the employer may not go ahead until a ruling has been given by an independent conciliation committee.

Breaks

Male employees must be given a break of at least half an hour in any working period lasting longer than six hours, or two breaks of fifteen minutes. Women have longer breaks. Young people must have a break of at least thirty minutes for any working period of between four and a half and six hours, and of sixty minutes where the working period is longer than six hours. Collective agreements can improve on the statutory minimum, and in some cases grant additional breaks for workers on moving assembly lines, or where VDUs are in use.

Protected groups

Night work is not customarily permitted for women classified as blue-collar workers, except in industries such as transport, entertainment, hotels, restaurants, the health services, and in markets.

Pregnant women and nursing mothers may not work overtime (in this case beyond eight and a half hours a day), work between 20.00 and 06.00, or work on Sundays. There are some exceptions in early pregnancy in hotels and restaurants, and weekend work is permitted in a number of service and caring occupations, as long as the women are guaranteed a twenty-four-hour break during the working week.

Concessions for special groups are also provided for in collective agreements: workers over 58 in the chemical industry, for example, work a thirty-five-hour week, compared to thirty-nine in the industry as a whole; employees in the chemical industry aged over 58 with ten years' service can also opt to work twenty hours a week for 70 per cent of their former gross earnings or — should part-time work be impossible — may be offered early retirement on 75 per cent of former pay.

Shift work

In establishments with a works council the introduction of shift work is subject to co-determination. The length of shifts and minimum rest periods (eleven hours between shifts) must be in accordance with the regulations set out in the Working Time Ordinance. Single-shift working is still very common in medium-sized firms in manufacturing, with two shifts in larger and more capital-intensive industries: round-the-clock working is uncommon, with the exception of continuous process industries. Here, principally in the chemical industry, some experiments have been under way with five-crew shift systems yielding a working week of 33·6 hours for each employee. At Bayer, where one trial has been running, some employees have been granted a thirty-seven-hour week — compared with the thirty-nine agreed in the industry — for working a five-crew system: the difference between the mathematical working week of 33·6 under five teams and the thirty-seven-hour week agreed for this group is made up for by the workers putting in an extra seventeen workdays per year.

Flexibility

Working time flexibility has been a major element in the process of cutting working hours. A number of new provisions were made in the industry agreements in metalworking in 1984, 1987 and 1990, when reductions were agreed. The essence of the 1984 and 1987 flexibility provisions concerned systems of averaging hours, with the possibility of limited plant, departmental, and individual variation in working hours within a reference period of, from 1987, six months. Under the 1990 agreement, which provides for the introduction of a thirty-five-hour week from 1 October 1995, the individual variation within a six-month period is retained, together with a new clause which allows up to 18 per cent of an establishment's work force to work up to forty hours on an individual basis, either for higher pay at normal rates, or for additional time off. Employees willing to work the longer week must commit themselves for two years. Although no empirical studies have been released showing whether the new opportunities have been made use of, trade union surveys of the implementation of flexibility in the 1980s showed that steps to exploit flexibility provisions were still tentative in all but the largest firms.

Time off

Annual leave

The legal annual minimum leave entitlement, under legislation dating from 1963, is eighteen days, based on a six-day working week, with a longer entitlement for young people under eighteen. In practice, leave for most employees is set by collective agreement. Under the law, and in most agreements, full leave entitlement begins after six months' service. Entitlement expires by 31 March the following year.

Most adult employees have six weeks' annual leave, in the vast majority of cases independent of age or length of service, although such provision continues to apply in industries such as wholesaling, road haulage and brewing.

Public holidays

There are ten national public holidays, with an additional four religious holidays which apply only in certain *Länder*. From 1991, what was German Unity Day (17 June) will be replaced by a holiday on 3 October to celebrate the act of unification between the Federal Republic and the German Democratic Republic in 1990.

Sick leave

Employees are entitled to full pay from their employer for the first six weeks of illness. This provision does not apply to blue-collar employees with less than one month's service, or to part-time blue-collar workers with a working week of less than ten hours a week or forty-five hours a month. Blue-collar workers must present a doctor's certificate after three days: in the case of white-collar workers no such requirement is imposed by law, although collective agreements, works agreements and individual contracts often require it.

After six weeks employees receive sickness benefit from their sickness insurance fund. This is equivalent to 80 per cent of previous regular earnings, subject to the contributions ceiling, but may not exceed previous net earnings. Benefit is paid for up to seventy-eight weeks within a three-year period for the same illness.

This provision was referred by the Federal Labour Court to the Constitutional Court for a ruling on its conformity with the Basic Law, as it constitutes discriminatory treatment between full- and part-timers. In 1989 the European Court of Justice ruled that the provision represented indirect sex discrimination, as most part-time workers are women. Some Social Democratic *Länder* have prepared draft legislation which may be introduced into the Upper House (the Bundesrat) in the coming legislative period to remedy this.

Maternity and parental leave

Female employees are entitled to six weeks off before the expected date of confinement and eight weeks after the delivery (lengthened to twelve weeks in the case of premature or multiple births). Maternity benefit is payable by the sickness insurance fund up to a maximum of DM 25 per day (£8.60), with the difference between this and previous net pay made up by the employer.

Either parent is currently entitled to eighteen months off for children born after 30 June 1990, once statutory maternity leave has expired. If one partner was not in regular work, or seeking work as a registered unemployed person, before the birth, and is able to look after the child, then the other partner is not entitled to leave. Parental benefit is payable at a rate of DM 600 (£206) per month up to the seventh month, when it is reduced according to household income. Parental leave will be extended to three years from 1 January 1992, but the current duration of parental benefit will be left unchanged.

Supplementary provisions for parental leave and career breaks have been made by many companies eager to recruit and retain women. In some cases they have simply taken the form of career breaks, with right of return to comparable work, which are longer than those provided by statutory parental leave. In other instances, employees are granted leave but encouraged to participate in training and retraining, and to work part-time or as temporary replacements. Some firms allow part of the career break to count towards eligible service for an occupational pension. The right of return is usually qualified, in some cases if there are redundancies, but some firms have agreed arbitration machinery to resolve disputes.

Family events

Paid leave for family and personal needs is provided in many collective agreements for events such as births, marriages and funerals, with, typically, one to three days for each event. Leave may be improved at company level, although not always on full pay: at Siemens, for example, employees may take up to one year's unpaid leave to care for a member of the family.

In the former GDR women working full-time will continue to have one day off per month up to the end of 1992 for housework if they are married, have children under 18 or members of the family who need care, or are over 40 years of age.

Educational leave

Educational leave, paid and usually totalling a week, has been legislated for in the majority of *Länder* in the former FRG. The Employees Continuing Education Act 1985 in North Rhine–Westphalia, for example, grants five days' leave per year for 'vocational and civic education' on full pay at recognized educational establishments.

Employee representation

Works councils, which may be elected by employees in any establishment with five or more employees, meet during working hours, and works councillors, directly elected by the work force of an establishment, are released from work for meetings. Some works councillors are released for works council duties full-time. This right does not extend to trade union officials, although some collective agreements give a right to attend union training courses.

Military service

Employees called up for military service continue in their previous employment and cannot normally be dismissed for absence for military service alone.

Retirement

Normal retirement age is 65 for both men and women. Retirement is not automatic at this age, although this is typically provided for in collective agreements, works agreements or individual agreements drawn up under company practice. Where this is the case, no notice to terminate the employment relationship is usually required.

Women may retire at 60 and draw a state pension if they have a minimum of fifteen years' insured employment and have paid pension contributions for at least ten out of the previous twenty years. Unemployed men may retire when they reach 60, provided they have fifteen years' insured employment, with pension contributions in eight out of the previous ten years. Men with thirty-five years' insured employment may retire at 63. Disabled employees can retire at 60, provided they have thirty-five years' insured employment.

Under 1989 legislation, employees who have reached 58 may apply to work half their normal working hours, but not less than eighteen per week, in return for 70 per cent of previous net pay. The employer pays 50 per cent of the previous pay, with a 20 per cent tax-free addition from the Federal Employment Institute — the German labour authorities — if the vacant post is occupied by an unemployed person. Some industries and individual companies operate comparable schemes, though without the state subsidy.

The Pension Reform Act 1989 will raise retirement ages as from 2001. Initially, existing early retirement options will be phased out by three months a year, and from 2004 by six months a year, to abolish these concessions by 2006. Women and the unemployed will have their retirement age raised to 65 by 2012. Pensions will become linked to the growth of net rather than gross earnings.

The state pension scheme, financed jointly by employer and employee contributions, pays on average 40–45 per cent of final earnings after a full career. Company schemes typically add to this to raise post-retirement income to between 60 and 70 per cent of final gross earnings. Many pension promises are met direct from book reserves, rather than funded, although supplementary insurance protects payments in the event of insolvency. Company pensions must be reviewed every three years.

Equal treatment

Sex equality

Under the Federal German basic law of 1949 no one may be disadvantaged because of sex, origin, race, language, home, creed or religious and political beliefs. As noted above, similar provisions apply through the Works Constitution Act, under which the works council's agreement is required for appointments.

Article 3(2) of the basic law states that 'Men and women shall have equal rights'. Federal Germany has also ratified ILO Conventions 100 and 111, which provide for equal pay and equal treatment at work. Under the European Community Adjustment Act 1980, the Federal German Civil Code was amended to incorporate EC directives on equal treatment. Article 611(a) of the Civil Code forbids discrimination against an employee in the sphere of appointments, promotion or dismissal, except where the gender of the prospective employee is directly relevant to the activity required and where a particular gender is an indispensable condition of a particular job. Should an employee be able to show that a disadvantage on grounds of sex may be taking place, the burden of proof rests with the employer that any unequal treatment is justified on material grounds not related to the sex of the employee or that a particular gender is an indispensable requirement of the job. Article 611(b) of the Civil Code requires employers to refrain from issuing job descriptions or job advertisements which specify gender, unless a particular gender is an indispensable prerequisite.

In 1984 the European Court of Justice ruled that the restitution of out-of-pocket expenses was incompatible with the EC's 1976 directive on equal treatment in that it did not constitute a sufficient or effective deterrent to employers. In the light of the Court's ruling, two Federal German labour courts reversed previous judgements which would have awarded two plaintiffs out-of-pocket expenses of DM 7.20 (£3.40) and DM 2.31 (£0.80) respectively, and instead awarded damages equivalent to six months' salary in the posts concerned, totalling DM 21,000 (£7,000) in each case, based on article 823 of the Civil Code ('Any person who deliberately or through neglect unlawfully harms the life, person, health, freedom or property or any other right of another is liable to pay compensation for any damage incurred').

Draft legislation was issued by the federal government in spring 1990 aimed at providing a new legislative basis for awards in the area of sex discrimination. The draft 'law to improve the equal treatment of men and women at the workplace' proposes maximum compensation of up to four months' pay for the post if a candidate is denied a job on the grounds of sex, reduced to three months if discrimination was not the ultimate reason why the applicant was rejected. Should several applications be discriminated against during a single selection procedure for a number of posts, then total compensation to all may not exceed ten months' pay, or five months if only one post is open. The same principles will apply to promotion, should the promotion itself not be possible for the applicant.

Legislation is also likely to be submitted by certain Social Democratic-ruled *Länder* to correct the situation of indirect discrimination caused by different qualification requirements for sick pay.

The disabled

Under the Severely Handicapped Persons Act 1986, all private and public employers with at least sixteen posts are required to employ handicapped persons in at least 6 per cent of those posts. 'Handicapped person' for the purpose of the Act is any person suffering at least a 50 per cent disability: persons with a 30–50 per cent disability may apply to the local employment office for treatment on the same footing as persons with a 50 per cent disability. Employers who fail to fill their prescribed number of reserved posts are obliged to pay a sum in compensation — currently DM 150 (£50) per month — for each unfilled place.

Work environment

Under Article 2 of the Basic Law, the state has a basic duty to protect the life and health of citizens. However, health and safety at work is not a centralized function in Germany, and there are two, in part overlapping, sets of authorities: the official Labour Inspectorate

(*Gewerbeaufsicht*), organized at Land level, and the non-governmental Accident Prevention Institutes (*Berufsgenossenschaften*). The latter provide accident insurance cover for companies, issue accident-prevention regulations, and are financed by insurance premiums levied on employers according to risk. In addition, the Federal Ministry of Labour also has a responsibility for issuing regulations, and oversees a programme intended to improve job design ('work humanization'). The Federal Institute for Occupational Safety, based in Dortmund, is the government research organization for safety and accident prevention. Both the Labour Inspectorate and Accident Prevention Institutes are entitled to issue regulations on safety and accident prevention which in general complement each other.

The most important general statutes regulating the work environment are:

- the Industrial Code (*Gewerbeordnung*).
- the Machine Safety Act (*Gerätesicherheitsgesetz*).
- the Imperial Insurance Ordinance (*Reichsversicherungsordnung*).
- the Law on Hazardous Materials (*Chemikaliengesetz*).

The Industrial Code sets out the basic health and safety obligations of the employer to protect the life and health of employees from hazards at work, insofar as this is permitted by the nature of the activity. The Code outlines, but does not itself detail, the minimum requirements for the work environment such as sufficient light and ventilation and measures to control dust, fumes and wastes. The Code is the statutory basis on which detailed regulations can be issued by the Factory Inspectorate. For the work environment the most important of these are the 1975 Workplace Regulations (*Arbeitsstättenverordnung*), which make detailed prescriptions on the size of working areas, lighting, ventilation, noise, internal transport, rest rooms and toilets for employees, and the maintenance of safety equipment.

The 1968 Machine Safety Act, together with regulations issued by the Accident Prevention Institutes, sets standards for the design, manufacture and operation of industrial equipment. The law obliges the manufacturer and importer of equipment to make available only products whose normal methods of operation (in accordance with manufacturers' instructions) are safe both for operators and third parties. Detailed provisions under the Act are regularly issued by the Minister of Labour and published in the Federal Employment Gazette (*Bundesarbeitsblatt*).

The Imperial Insurance Ordinance is the statutory basis for the activities of the Accident Prevention Institutes, which issue detailed regulations complementing the provisions issued by the Ministry, and in many instances setting standards valid beyond their immediate members.

The Law on Hazardous Materials is the primary legislation on the manufacture and use of materials, with special reference to materials used at the work place. It provides for the classification of materials, testing requirements, and notification obligations. It constitutes the statutory basis for the Hazardous Materials Regulations (*Gefahrenstoffverordnung*) which incorporate EC law in this area into German law and set out detailed regulations on the introduction and use of hazardous materials.

Work place safety organization

German law imposes a number of requirements for the appointment of safety specialists and company physicians, and also gives Works Councils rights and obligations in the field of health and safety.

Under the 1973 Safety at Work Act (*Arbeitssicherheitsgesetz*), employers must appoint safety engineers, physicians and other safety specialists; which establishments are required to do this, and how, is determined by size and degree of risk according to detailed regulations issued under the Act. (The average threshold is 50 employees.) A number of small or medium-sized firms typically share one physician through a joint-firm medical centre. In addition, the Insurance Ordinance requires that any undertaking with more than 20 employees must appoint a safety officer (*Sicherheitsbeauftragte*), with specific responsibilities for accident prevention.

Works Councils have a number of rights and duties in the health and safety field. These include:

- A right of codetermination over measures designed to prevent accidents and occupational disease.
- A right to be consulted over the appointment of safety officers, and to approve the appointment of safety specialists and physicians.

- A duty to ensure the implementation of law, ordinances and safety regulations, and specifically to support the Accident Prevention Institutes and other 'competent bodies' in eliminating hazards.
- A right to participate in inspections organized by the employer.
- A right to be informed over plans concerning premises, equipment, operations and jobs, and to make suggestions as to how the 'established findings of ergonomics' can assist in tailoring jobs to human requirements.

Termination of contract

Law and practice on the termination of a contract in Germany differ in a number of respects from UK law. Paramountly, there is a legal requirement on employers to notify works councils, which may be established in establishments with five or more employees, of any proposed dismissal, whether 'ordinary' — that is, with notice — or summary. Any termination not notified to the works council, setting out the reasons, is null and void. Termination of executives' appointments must also be notified to the elected 'executive representation committee' (*Sprecherausschuß*). Failure to do so renders the termination null and void. In all instances of dismissal German law requires the termination of employment to be the measure of last resort.

Although notice to terminate does not require a particular form, this may be specified in a collective agreement or individual contract. Termination of a training agreement must be in writing. Failure to use the required form will render the proposed termination invalid. Notice does not have to state the reasons for the dismissal, except in the case of trainees. However, the Civil Code requires that reasons for the dismissal must be given on request in the case of a summary dismissal. In addition, any notification to a works council must include the reasons for the dismissal. Where termination is on economic grounds the employer must, if asked, inform the employee of the reasons for the choice of employees for dismissal.

Dismissal on non-economic grounds

Depending on the reasons which prompt the dismissal, a contract of employment may be terminated either by 'ordinary dismissal' with notice or by summary dismissal.

Employees aged over 18 with at least six months' service and employed in an establishment with at least six employees are protected against unfair ('socially unwarranted') dismissal under the Protection against Dismissal Act 1969. In calculating the number of employees in the establishment, only those working for more than ten hours per week or forty-five hours per month count. Trainees are excluded.

A 'socially unwarranted' dismissal is null and void. An 'ordinary' dismissal — that is, a dismissal with notice — is warranted only if it is prompted by the employee's person or conduct, or by 'urgent operational reasons' (see below). A dismissal will be unfair if it contravenes any anti-discrimination provision set out in the Works Constitution Act or if the employee could be employed elsewhere in the firm, and the works council objects to dismissal.

In general the longer the period of employment, the more serious the reasons warranting dismissal have to be. Personal factors which could justify an ordinary dismissal may include:

- Basic unsuitability, lack of skills and inability to learn them.
- Illness, if persistent or frequent with a high probability that it will not be cured. The Federal Labour Court has indicated that a total absence of at least 30 per cent over two successive calendar years would have to be shown.
- Age: reaching 65 — the normal age of retirement — is not itself ground for dismissal, and employment contracts need to specify that employment will end at 65 if this is desired. A number of collective agreements also protect employees above a certain age (usually 50 or 55) from either ordinary dismissal or redundancy.

Employee conduct which breaches a term of the contract of employment may warrant an ordinary dismissal. It might include: refusal to carry out duties, unexcused absence, leave without permission, unpunctuality, breaking company rules on alcohol or smoking, and aggressive or threatening behaviour. Breaches of broader contractual obligations, such as confidentiality, could also be cited. The employer must first warn an employee in writing that his or her conduct risks dismissal within a reasonable period from the alleged breach — four weeks would be the maximum. Employees can add their version of events to their file.

Summary dismissal. An employer may terminate an employee's contract without notice for 'an important reason', defined as 'circumstances under which it would be unreasonable . . . for the party terminating the contract that the contract of employment should continue until the expiry of the notice period'. Such serious breaches of contract might include criminal behaviour or persistent refusal to comply with the contract of employment. A summary dismissal must be effected within two weeks of the occurrence which has prompted the dismissal. The employer must inform the employee of the reason(s) for dismissal in writing on request. Summary dismissals have to be notified to works councils.

Works council objections. Both ordinary and summary dismissals must be notified to the works council, which has one week to respond in the event of an ordinary dismissal and three days in the event of a summary dismissal. Its response can take three forms:

- Assent to the dismissal.
- Objection: grounds might include failure by the employer to warn the employee properly, for example, or that the penalty of dismissal is inappropriate for the stated conduct.
- In the case of an ordinary dismissal only, an objection is possible on the grounds set out in section 102, para. 3, of the Works Councils Act, which imposes limitations on employer decisions, most notably in the event of redundancies: these grounds are set out below in the section dealing with redundancy.

A works council objection does not prevent the employer from issuing notice to the employee. However, any objections must be appended when notice is sent to the employee. If the works council objects to an ordinary dismissal and the employee, if entitled to do so, has initiated proceedings at a labour court within three weeks of the notice being given, then the employer must continue to employ that person if the employee requests it until settlement of the case (see below).

Dismissal on economic grounds

A dismissal on economic grounds is considered fair if 'urgent operating reasons' warrant it and if the works council does not object. 'Urgent operating reasons' would include factors outside the control of the establishment (such as a fall in orders) or proposed changes in the establishment's work organization (rationalization, closure of all or parts of an establishment) which mean that employment cannot be sustained at its existing level.

Employers are free to determine the measures that are needed to meet 'urgent operating exigencies', as long as their decision is not 'patently irrelevant, unreasonable or arbitrary'. Employers must also examine whether redundancies could be avoided by short time, less overtime, internal transfer of employees, or other measures. Employers must ensure that the choice of employee meets a set of 'social criteria'. Employees dismissed must be those least 'socially' affected. Factors to be taken into account include: length of service, age, family status and the number and incomes of dependants, health, occupational disability, membership of a protected group. Works councils and employers can agree a points system for redundancies in an establishment. Performance, in general, may not play a role in selection for redundancy.

Employers can plead against the application of 'social criteria' where it is justified by 'operational, economic or other legitimate requirements' related to the establishment. On request, the employer is obliged to inform a dismissed employee of the reasons for the 'social selection'. This can entail revealing the other employees included in the selection process by name.

An employer must also refrain from dismissing an employee if their continued employment would be possible after a reasonable period of training or retraining or if their employment might be possible under changed conditions of employment. A notice to change the contract is subject to the same notice periods as other forms of termination, and can also be contested by the employee in the courts.

Notification requirements for redundancies. Notification and consultation procedures depend on the number of proposed redundancies and the size of the establishment. The procedure is basically the same as for a dismissal on non-economic grounds. The Works Constitution Act also imposes a number of criteria which a works council can cite in objecting to a dismissal, most of which are relevant to redundancies rather than disciplinary dismissals. These are:

- Failure to observe social criteria in the choice of employee(s).
- Contravention of a guideline for selection provided for under section 95 of the Works Constitution Act. (In establishments employing more than 1,000 employees the works council can demand the setting up of 'selection criteria' for appointments, transfers and dismissals. In establishments with less than this number of employees any criteria proposed by the employer require the agreement of the works council.)
- The possibility that the employee can be employed elsewhere in the establishment or company, if necessary after a reasonable period of training.
- The possibility of continuing employment under changed contractual conditions where the employee is willing.

Redundancy pay. There is no legal obligation on employers to make payments for redundancy over and above normal pay during notice. However, provision for a redundancy payment may be included in a collective agreement, an individual contract or a works agreement as part of a so-called 'Social Plan': the latter are dealt with in the context of collective dismissals in volume 3 of the European Management Guide series.

Where a severance payment is provided for, or granted for an individual redundancy, it is often calculated on the same basis as compensation for unfair dismissal: that is, one month's salary for each year of service up to twelve years' (more for older employees).

Redundancy pay — collective agreements

Many collective agreements contain clauses providing for earnings guarantees and redundancy protection for older employees, and compensatory payments for employees. For example, the framework agreement in the chemical industry provides for severance compensation, following redundancies caused by rationalization, ranging from one month's pay for a 40 year old with 10 years' service to six months' pay for a 60 year old with twenty-five years' service.

Terminating executives' employment

Executives (*leitende Angestellte*) fall under the scope of dismissal protection legislation but with a number of important limitations. First, the Protection against Dismissal Act does not apply to persons who are members of the body which legally represents a company, usually and in practice the management board (*Vorstand*). Second, executives, heads of department and 'similar senior white-collar employees' do not have the right to take their objection to what they feel is a socially unwarranted dismissal to the works council. Employers are required only to *inform* the works council of notice to terminate such an executive's contract. A dismissal will still be valid even if the works council is not consulted. An employer is not obliged to give the reasons for not reinstating an unfairly dismissed executive. That is, a court must comply with an employer's application to terminate the employment relation even when the employee has been unfairly dismissed.

Any Executive Representation Committee (*Sprecherausschuß*) elected under legislation passed in 1988 must be informed, with reasons given for the termination. Any termination of an executive's employment without consulting the committee will be null and void. The committee has a week to respond (three days in the case of summary dismissal).

Period of notice

During a probationary period, the contract may be terminated in accordance with the legal minimum notice: that is, two weeks for a blue-collar worker and four weeks for a white-collar employee.

Long-service employees are entitled to longer periods of notice, although there are differences between blue and white-collar provision, with white-collar employees still enjoying longer statutory minimum periods of notice than blue-collar workers. Legislation setting a common qualifying age from which service must be counted for the calculation of longer notice periods for both groups was passed in May 1990. Service after the age of 25 now counts for white and blue-collar employees, whereas previously blue-collar workers could count service from only the age of 35. In June 1990 the Federal Constitutional Court ruled that legislation to equalize the actual length of notice periods must be introduced by 30 June 1993. Controversy can be expected over whether this will entail lengthening blue-collar notice or shortening white-collar provision.

In practice, many collective agreements have already abolished the difference, and brought blue-collar employees up to the white-collar level. Moreover, and provided it is done by collective agreement, shorter periods of notice than the statutory period may be agreed.

In the case of *blue-collar* workers the statutory minimum periods of notice for termination by the employer are laid down in section 622, para. 2, of the Civil Code. They are:

- For all workers, two weeks.
- For workers with five to nine years' service, one month.
- For workers with ten to nineteen years' service, two months.
- For workers with twenty or more years' service, three months.

Under the 1990 law, years of service over the age of 25 are taken into account for the purposes of calculating periods of notice.

Notice must be given such that employment is terminated on the last day of a calendar month (unless the employee is entitled to three months' notice, in which case it takes effect from the last day of a calendar quarter). For example, an employee with three years' service would have to be given notice on 17 January for employment to be terminated on 31 January. In the case of an employee aged over 55, with twenty years' service, notice could be given only on 31 December, 31 March, 30 June or 30 September to terminate employment on 31 March, 30 June, 30 September or 31 December. This timetable does *not* take account of the obligation to consult the works council and wait for its reaction. Thus, if notice is to be given on 31 March, the works council would have to be notified by 24 March, and so on.

Statutory minimum notice periods of notice for an employer wishing to dismiss a *white-collar* worker are laid down in section 622, para. 1, of the Civil Code. They are:

- For all employees, normally six weeks but in no case less than one month.
- For employees with over five years' service, three months.
- For employees with over eight years' service, four months.
- For employees with over ten years' service, five months.
- For employees with over twelve years' service, six months.

Only years of service over the age of 25 are taken into account in calculating periods of notice.

Notice must be given such that employment is terminated on the last day of a calendar quarter in all cases. The last days in each case when notice may be given are therefore 17 February, 19 May, 19 August and 19 November.

In the former GDR previous legislation remains in force. It provides for minimum notice of two weeks for all employees, rising to one month after five years' service, two months after ten years' service, and three months after twenty years' service. These periods may be increased by collective agreement, and only service after the age of 25 is taken into account.

Termination by mutual agreement

An employee's contract may be terminated by mutual agreement, with the payment of a sum in compensation to the employee. This can have tax advantages for the employee, as severance payments up to certain limits are free of income tax. The limits are: DM 24,000 (£8,250) for all employees; DM 30,000 (£10,310) for employees aged at least 50 with a minimum of fifteen years' service; DM 36,000 (£12,415) for employees aged at least 55 with twenty years' service. Severance payments above these limits are subject to income tax, but the rate can be cut to 50 per cent of the nominal rate payable on application to the tax authorities. Severance compensation is also offset against unemployment benefit.

Such an agreement will normally specify: when the employment will end and the reason; any compensation; provision for time off to find another job; any restrictive covenants, and a 'settlement clause' under which the employee discharges the employer from all obligations under the contract of employment. One important element is that the employee agrees to allow employment to end before the expiry of the notice period stipulated in law, collective agreement or individual contract.

Protected employees

Certain categories of employee are protected both against ordinary and, under certain circumstances, against summary dismissal.

- *Conscripts* on military service are protected against ordinary dismissal but not against summary dismissal.
- *Pregnant women* are protected against both ordinary and summary dismissal during pregnancy, provided that the woman has informed her employer of her pregnancy or does so within two weeks of notice being given. She is also protected for four months after confinement.

- *Parents on parental leave.* A parent on parental leave cannot be dismissed. The right can be removed with the permission of the *Land* labour office under exceptional circumstances, such as the closure of the business and the impossibility of finding alternative employment.

  In the former GDR lone parents of children under 3 years old are protected from dismissal if the child is born before 1 January 1992. (This may be relaxed in the event of the partial or total closure of the establishment.)
- *Disabled employees.* Ordinary or summary dismissal of a disabled employee requires prior notification in writing to the Central Disabled Workers' Office and its consent.
- *Apprentices/trainees.* During probationary periods, which must be no shorter than one month and no longer than three months, an apprenticeship or training contract may be terminated at any time without notice. After that period a training contract may be terminated only for an 'important reason'.
- *Members of works councils* are protected against ordinary dismissal during their office and for one year thereafter. Summary dismissal of such employees is subject to works council approval. If it is refused, but employers still wish to dismiss, they must apply for a court ruling. Members of the electoral board (which administers works council elections) are protected against ordinary dismissal during office and for six months thereafter. Candidates for election to a works council are protected against ordinary dismissal from their nomination until six months after publication of the result.
- '*Older employees*'. In many collective agreements older employees (usually over 53–55) with a minimum qualifying period of service (usually three years) can be dismissed only through a summary dismissal for an 'important reason'.

Unfair dismissal procedure

The Protection against Dismissal Act, which defines a 'socially unwarranted' — 'unfair' to borrow English terminology — dismissal and sets out appeal procedures, applies only:

- In establishments employing six or more employees, excluding trainees.
- To employees with at least six months' service.

The grounds for deeming a dismissal to be unwarranted — that is, if it is not occasioned by the person or conduct of the employee, or urgent operating requirements — have already been outlined above.

An employee who wants to object to notice of dismissal must notify the works council within a week. If the works council accepts the objection it is obliged to seek a resolution of the differences between employer and employee. If no resolution is achieved, and the employee wishes to contest the dismissal, the employee must appeal to a labour court within three weeks of the notice of dismissal, attaching the opinion of the works council. The burden of proof rests with the employer to demonstrate that the dismissal was socially warranted.

The employee has a right to continued employment — that is, work at the work place — until final settlement of the case if the works council objected to an ordinary dismissal and if the employee has initiated proceedings before a labour court. The employer may be exempted if the action brought by the employee is unlikely to succeed or appears abusive, if the continuation of the employment would constitute an unreasonable burden on the employer, or if the works council's objection was 'patently unfounded'. However, between the expiry of the period of notice and the first instance hearing of a labour court the balance of 'justified interest' in general lies with the employer: the employer does not have to actually employ the complainant whilst waiting for the ruling. However, should the labour court of first instance rule in favour of the employee, and the employer appeals, then the balance of justified interest is held to move to the employee. The complainant must be employed, that is, offered actual work, between the first and second-instance hearings.

Reinstatement and compensation

If a labour court upholds the appeal, and the employer takes no further steps to try and reverse the ruling, the employee has a right to reinstatement on the former conditions of employment and to full pay for the period between the expiry of the notice period and the date of reinstatement. The employer may deduct any earnings actually earned in the intervening period or which might have been earned had the employee not wilfully refused other employment and any state social benefits.

Reinstatement is rare, however. Either party can apply to the court to terminate the employment relation in return for a compensation payment to the employee. Under the Protection against Dismissal Act, the maximum compensation is twelve months' pay, which can be increased to fifteen times monthly pay for employees aged 50 with at least fifteen years' service, and up to eighteen times monthly pay for employees aged 55 and over with at least twenty years' service.

Final formalities

If the dismissal is not contested, on expiry of the legal or contractual period of notice the employer must give the employee all the documents relevant to the employment, completed and up to date: that is, tax documents, social security record, a final pay statement, details of holiday taken, etc. The employee may also require the employer to provide a written certificate (*Zeugnis*) setting out the type and length of the employment including, if requested, details of the employee's conduct and performance. Case law has determined that such a certificate must be true yet must not unnecessarily obstruct the professional career of the outgoing employee. The information contained in it must only be that in which a future employer has a legitimate interest. An employee has the right to demand, if necessary through the courts, that a new leaving certificate is issued in the event of inaccuracies.

Claims to outstanding pay or other forms of financial compensation lapse two years after the calendar year in which entitlement was earned. Collective agreements may also include provision for a lapsing of claims arising out of the employment relationship if they are not lodged within an agreed period, customarily up to six months.

Appendix

Federal Ministry of Labour
(Bundesministerium für Arbeit
und Sozialordnung)
5300 Bonn 1
Postfach 14 02 80
Rochustrasse 1
tel. +49 228 5271
fax. +49 228 527 2965

Federal Employment Institute
(Bundesanstalt für Arbeit, BfA)
Regensburger Strasse 104
8500 Nürnberg 30
tel. +49 911 170
fax +49 911 17 21 23

The Bundesanstalt operates regionally through nine state employment offices (*Landesarbeitsämter*) in each constituent state of the republic, and 150 local employment offices.

Federal Association of German
Management Consultants
(Bundesverband Deutscher
Unternehmensberater, BDU)
Friedrich-Wilhelm-Strasse 2
5300 Bonn 1
tel. +49 228 23 80 55
fax +49 228 23 06 25

British Chamber of Commerce in
Germany
Heumarkt 14
5000 Köln 1
tel. +49 221 234284

Confederation of German
Employers' Associations
(Bundesvereinigung der
deutschen Arbeitgeberverbände,
BDA)
Gustav Heinemann Ufer 72
5000 Köln 51
tel. +49 221 37950
fax +49 221 37 95 235

German Personnel Management
Society (Deutsche Gesellschaft
für Personalführung, DGFP)
4000 Düsseldorf 1
Niederkasseler Lohweg 16
tel. +49 211 59 78 0
fax +49 211 59 78 505

German Trade Union
Confederation (Deutscher
Gewerkschaftsbund, DGB)
Hans-Böckler-Haus
Hans-Böckler-Strasse 39
4000 Düsseldorf 30
tel. +49 211 43011
fax. +49 211 4301471

The Federal Institute for
Occupational Safety
(Bundesanstalt für Arbeitsschutz)
4600 Dortmund 17
Postfach 17 02 22
Vogelpothsweg 50-52
tel. +49 231 1 76 31
fax. +49 231 176 3454

5

Greece

Terms and conditions of employment in Greece are regulated both by statute and by case law, which can prove decisive in determining the outcome of a contested dismissal. The main pieces of legislation governing individual terms and conditions are the Civil Code, laws 2112/1920 and 3198/1955 on termination, and laws 4504/1966 and 1346/1983, which regulate annual leave. Law 1483/1984 makes provision for leave for employees with family responsibilities.

Minimum pay and normal working hours are settled by central collective bargaining, with scope for individual employers to improve on the agreed provisions. The terms of these central agreements are binding on all employers following a declaration to that effect by the Ministry of Labour.

Contracts of employment

Contracts of employment may be, but do not have to be, in writing, except where the contract is for a fixed term (see below). An oral agreement is acceptable for an open-ended contract of employment. In practice written contracts are a rarity and more or less exclusively confined to managerial employees. Where the contract might involve a special condition, going beyond the normally accepted terms of employment, companies — and employees — may seek to protect themselves by insisting on a contract, or agreed letter of engagement. This might apply, for example, where a company might want to send sales employees for long periods of travel, or transfer staff out of the Athens area — widely seen as a disbenefit — and wished to ensure that there was evidence that employees had accepted the conditions prior to engagement.

From the point of view of senior employees, a contract might also be signed under which a new employer would agree to pay a higher than statutory severance indemnity for an initial period of employment — typically, three years — to reflect the fact that a manager had accumulated several years of service-related but non-transferable indemnity with a previous employer. (Provisions on service indemnity are set out in the section on individual termination, below.) Management trainees might also be given a formal letter of engagement.

Larger — and in particular multinational — companies will often use a simple 'form of engagement' for all grades of employees: the same form will serve record-keeping purposes, as well as being proof — if needed — of the basic terms of employment and pay. Such a form typically records the date of engagement, the department, the employee's function, the pay, any transfers to other departments, and details of dismissal or resignation.

In the event of a dispute between the parties as to the substance of a contract, the burden of proof lies with the party seeking to obtain their rights. What counts as the substance will be, in general, the basic facts of the employment relation: pay, hours, other benefits. Most of these elements are either derived from collective agreements or are a matter of record (from the tax and social insurance authorities, for example).

A contract of employment with a minor below the age specified for that employment will be null and void, irrespective of parental consent (Civil Code, articles 127 and 136).

Any term in a collective agreement or individual contract of employment which conflicts with the provisions of the 1984 law on sex equality will be null and void (see below).

Probationary period

A probationary period can be agreed, although it is not common practice for most ordinary blue and white-collar employees. Two methods are possible. On the one hand, the two parties may agree a fixed-term contract for a period of, say, three or six months, after which a permanent post will be offered, subject to satisfactory performance (see below). Alternatively the employee can be hired immediately on an indefinite contract and performance reviewed after two months — after which employee liability for severance indemnity begins.

Employees are categorized as either blue-collar or clerical on the basis of their tasks: blue-collar workers cannot be asked to carry out clerical or administrative work, otherwise they must be reclassified as clerical employees. Some multinational employers have classified all their employees as clerical, both to facilitate flexibility and as a form of benefit: the main practical difference between the two categories is the fact that clerical employees enjoy a substantially larger severance indemnity on dismissal or retirement (see below).

Restrictive covenants

Restrictive covenants, imposing constraints on an employee's employment after resignation, would not be recognized by the courts and are therefore of little juridical importance. However, a company operating in a specialized area of the labour market might ask an employee to agree, in writing, to a provision under which they would undertake not to work for a direct competitor or set up in competition for an agreed period. Although ultimately unenforceable, the agreement would have a moral and practical force for executives interested in preserving their reputation.

Change of terms

Changing the terms of employment can be problematic where the change involves not necessarily or solely pay or material benefits, but also any diminution of the employee's status within an enterprise. Such a change, without the employee's agreement, can be contested in the courts. (This does not detract from the employer's ultimate ability to terminate an employee's contract on grounds of reorganization, should the employee continue to refuse the proposed change — see below.)

Fixed-term contracts

Fixed-term contracts may be agreed, either for short-term or for defined tasks — such as seasonal work or a sales campaign — or as a means of regulating a probationary period. Such contracts need to be in writing, and must specify the proposed period, pay and working hours. A fixed-term contract can be renewed once without difficulty, but a further renewal could lead the courts to deem that the employment relationship has become indefinite. This is not regulated by statute, and whether a second renewal would be acceptable would depend on the specific circumstances.

Working time

Hours of work

The legal maximum working week is forty-eight hours — based on six-day working — with a daily maximum of nine hours. In practice, most of industry works a forty-hour week — in line with national collective agreements, with thirty-eight or thirty-nine hours in some service industries, and thirty-seven hours in banks. Under the 1990 Development Law (law 1892) the normal working week can be extended to forty-eight hours for a period of three months, provided it is allowed for in a collective agreement. This law also allows employees to be hired to work two twelve-hour shifts at weekends only, with social insurance cover provided as if for full-time employees.

Double-day shift working is quite common in industry. A number of (mainly multinational) firms operate flexible starting and finishing times, but the practice of carrying over hours' debits or credits in a full flexi-time system is limited. Very few firms vary working hours according to the time of the year. In general there is no tradition of taking long lunchtime breaks or a siesta. Lunch breaks are often paid, or at least part paid, by companies — but tend to be very short by the standards of many European companies, typically ranging from twelve to thirty minutes a day.

Overtime

Any hours worked between the agreed forty and the permitted maximum of forty-eight per week are termed 'excess' or 'supplementary working', and must be paid at time-and-a-quarter. Based on five-day working, this would yield a practical limit to 'supplementary working' of five hours a week. True 'overtime' — that is, over and above forty-eight hours — requires the permission of the Ministry of Labour, and the supplements payable are on a sliding scale set out in law. If permission is not sought from the Ministry, overtime is deemed to be technically 'illegal', and payment increases to double time. In practice, many companies simply choose to pay double time in principle for all 'excess' and 'overtime' hours in order to avoid the formalities and to obtain flexibility at enterprise level.

The Ministry of Labour also sets annual overtime limits by occupation and industry, and these are reviewed every six months. Examples include a 120 hour limit for drivers, and seventy hours in manufacturing industry.

Daily and weekly limits may be exceeded in industry in cases of emergency, to carry out urgent work, to meet social needs, repairs and transport, or where necessitated by an act of God.

Night work

Night work, that is, work between 22.00 and 06.00, must be paid at time-and-a-quarter. It is prohibited for women and for young people up to 17 years of age in industrial enterprises. Night work on Sundays or at weekends is payable at double time.

Time off

Annual leave

A salaried employee with one year of service is entitled to annual paid leave of twenty working days. Leave is increased by one working day for each year of service up to a maximum of twenty-two days. This entitlement is often increased in company agreements or by unilateral company decision, with twenty-two days as a basis for all employees not untypical, with additional holiday by service and grade. The majority of larger companies in manufacturing industry have an annual shut-down.

Leave cannot be transferred to the next calendar year, and it is generally expected that half of all leave should be taken between 1 May and 30 September.

Public holidays

Obligatory public holidays consist of Christmas Day, 25 March, the second day of Easter and 15 August. In addition there are voluntary public holidays on 28 October and 1 May. Banks and government departments also take New Year's Day, Epiphany, the first Monday in Lent, Good Friday, the Day of the Holy Spirit and Boxing Day.

Maternity leave

Statutory maternity leave is sixteen weeks, fourteen of which are paid by the state at a rate of 50 per cent of normal earnings, with the balance made up by the employer. In practice, larger companies tend to pay the additional two weeks, to sustain earnings throughout the period of leave. The leave must be taken so that seven or eight weeks are before confinement and seven or eight after.

Parental leave

Under law 1483/1984 parents of either sex with responsibility for dependent children may apply for a period of unpaid parental leave lasting no longer than three months at any one time, subject to one year's service: the provision applies only in companies employing at least 100 workers, and no more than 8 per cent of a company's work force are entitled to the leave. The parent not applying for leave must also be working outside the household.

Leave may be taken following on from maternity leave up to thirty months from the birth. During the period of leave, employees continue to be responsible for paying social insurance contributions to the national fund (IKA). Workers are entitled to return to the same or a similar position.

Leave for workers with family responsibilities

Also under law 1483, employees in the private sector with dependent children are entitled to apply for unpaid leave of up to six days per year to look after a sick child or spouse. The leave may be extended for up to ten working days where the employee has more than two dependent children.

Parents of children with a chronic disability working in enterprises employing at least fifty workers may also apply for a one hour per day reduction in their normal working hours, subject to a corresponding reduction in pay.

Larger companies may also provide, either through agreement or on a unilateral basis, leave for other family events, such as the death of a close relative.

Sick leave

Absence through certificated illness for a relatively short period of time (specified in article 3 of law 4558/1930) is permissible without prejudice to the employee's contract of employment. Sickness benefit is paid out under the state scheme (IKA) for employees with qualifying service equal to 50 per cent of normal earnings, plus ten per cent of this sum for each dependant, up to a maximum of four. Benefit is reduced by one-third where there are no dependants. Benefit is initially payable for 180 days for the same illness, rising to 720 days, depending on the length of qualifying service. Under article 658 of the Civil Code the employer must pay the difference between the daily rate of the state benefit and the usual daily wage for a period of fifteen days for employees with more than ten days' but less than one year's service, and up to twenty-six days for employees with at least one year's service. There are three waiting days before the state benefit is paid, although the employer is obliged to pay half the daily wage during that period.

Educational leave

Employees aged under 25 undertaking educational courses may take fourteen days' leave a year in order to sit examinations. A benefit for this period is payable by the national employment organization, OAED.

Trade unionists

Under law 1264/1982 employers must grant unpaid leave of absence to any employee who is a member of the executive board of a trade union, or a delegate to a different tier of trade union, to enable them to carry out their duties. (The structures of trade unions, the law on industrial action and collective bargaining will be dealt with in volumes 3 and 4 of the European Management Guides series).

The amount of leave to be granted varies according to the size and status of the trade union. For example, for officials of representative national trade union organizations (the so-called 'third tier'), leave of absence must extend for the whole period of their office. This is reduced to five days for officials of 'primary' trade unions — that is, individual enterprise organizations — in companies with 500 or more employees, and to three days where they have fewer than

500 employees. Delegates are simply allowed time off to attend meetings.

Trade union leave is regarded as employment for the purpose of qualifying for all entitlements arising out of employment, or for fulfilling the qualifying period for benefits paid under social insurance. (However, the trade union itself is responsible for covering the social security contributions of any of its officials on leave of absence.)

Retirement

Normal retirement age under the state scheme (IKA) is 65 for men and 60 for women, with a minimum of thirteen and a half qualifying years (4,050 days). Employees are entitled to a severance indemnity on retirement which is equal to 50 per cent of the indemnity had the employee's contract been terminated by dismissal or redundancy. This is reduced to 40 per cent if the employee is covered by an auxiliary occupational scheme. (The calculation of the indemnity for blue-collar and white-collar employees is set out in the section dealing with individual termination of contract, below.)

In practice companies often pay an enhanced indemnity on retirement, either on an individual basis or as part of a benefits plan. In a 1989 survey of benefits practice carried out by the Greek Personnel Management Association (see appendix), thirty-seven out of sixty-nine surveyed companies paid extra indemnity on retirement, twenty of which paid double the statutory requirement. Thirty-three companies out of the seventy-nine provided auxiliary pension plans: the most common financing method was a deposit administration contract with an insurance company, with a guaranteed return plus the possibility of bonuses, depending on investment performance.

Pensions under the state scheme are based on the insurance class to which the employee belongs: there are twenty-two insurance classes, based on earnings. Basic benefit level for those with a minimum of ten years' service ranges from 30 per cent of earnings in the higher-income classes to 70 per cent for the lower classes. There are additions for long service, with 1 per cent of pay for the relevant insurance class added for each 300 contribution days, up to a maximum of 7,799 (approximately fifteen years) and a further 1·5–2·5 per cent of presumed pay addition for service beyond this period. There is a statutory auxiliary fund, TEAM, for employees not in a company scheme.

Early retirement up to three years before normal retirement age without any reduction in pension is permitted under the state scheme for employees with thirty-three and a third qualifying years of service (10,000 days): retirement at age 58 is possible for employees with thirty-five years' service (10,500 days). Earlier retirement options (from the age of 56 with thirty-five years' service) are also possible, subject to 10,500 days' qualifying service, with a cut in benefit of 0·5 per cent for each month of early retirement.

Equal treatment

Sex equality in employment is guaranteed under the 1975 constitution (article 22) and regulated by law 1414/1984. The law makes it unlawful to discriminate on grounds of sex or family situation in the areas of:

- Vocational guidance and training (except where positive action in favour of women is concerned).
- Access to employment.
- Remuneration (defined as wages plus other additional benefits in cash or kind, directly or indirectly for work performed).
- Conditions of work and professional development.
- Termination of contract (with protection against an employer terminating the employment of an employee who has sought to enforce his or her rights under the Act).

Trade unions have the right to inform employees of their rights under the Act, and employers must facilitate this, if requested, by allowing the use of company notice-boards and distribution of pamphlets (provided it is done outside working hours).

The Act provided for the establishment of a Sex Equality Section at the Ministry of Labour, whose tasks include collecting statistics on female employees and ensuring the application of measures to promote equality and remove inequality.

Penalties for infringing the Act range from Dr 20,000 to Dr 300,000 (£65–£970), payable to the Workers' Hearth social insurance fund. Employees may also resort to the civil courts in the event of unfair dismissal on grounds of sex, and under law 1264/82 trade unions may act on behalf of their members in such procedures.

The General Secretariat for Equality, established under law 1288/1982, with its responsibilities enhanced under law 1558/1985, is an autonomous department within the Ministry to the Presidency entrusted with a broad responsibility 'to promote and achieve the legal and substantive equality of men and women in Greece'. The secretariat can provide information on equality initiatives and company positive action programmes proposed under European Community recommendations.

Work environment

Law 1568/85 provides for the establishment of health and safety committees in all firms employing more than twenty people. The committees are responsible for assessing hazards, making proposals to improve them, monitoring compliance with detailed safety regulations, and generally overseeing safety and accident prevention policies in the enterprise. Where firms employ more than fifty employees a safety technician and factory doctor must be employed.

The Labour Inspectorate is under the responsibility of the Directorate for Organization of the Inspectorate within the Ministry of Labour (see Appendix). The Ministry also has a separate Directorate overseeing legislation and administrative controls on occupational hazards. Information on occupational medicine and health and safety issues can be obtained from the Centre for Occupational Health and Safety (see Appendix).

Individual termination of contract

Individual termination of contract does not require any distinction to be drawn between dismissals for factors related to the employee's person or conduct and dismissals on economic grounds or for purposes of reorganization and rationalization. Although Greek law makes provision for periods of notice, broadly speaking, employees can be dismissed without notice in return for the payment of an indemnity or with the period of notice set by the law for a reduced indemnity. The length of notice and levels of severance indemnity are set out below. Severance indemnity is payable in all instances of dismissal, including poor performance, except where the employee is potentially guilty of criminal misconduct and the employer chooses to take the employee to court.

Notice periods and levels of indemnity vary, depending on whether the employee is white-collar or blue-collar.

Dismissal of *white-collar* employees is regulated by law 2112/1920. Under its provisions an employer can terminate a contract of employment subject to the following periods of notice and/or levels of indemnity:

- Two months to one year's service: one month's notice or salary.
- One to four years: two months' notice or salary.
- Four to six years: three months' notice or salary.
- Six to eight years: four months' notice or salary.
- Eight to ten years: five months' notice or salary.
- Ten to twenty-eight years: an additional month's notice or salary up to a maximum of twenty-four months.

The full amount of indemnity is payable where the employee is dismissed without notice, and half the stated indemnity where the appropriate notice is given. If the amount of indemnity payable exceeds six months' salary, the employer may pay any sums in excess of the six months in three-monthly instalments, each of which may not be less than three months' pay. Failure to pay the indemnity or an instalment may render the termination invalid. The amount of indemnity is boosted by Christmas, Easter and holiday bonuses and calculated by multiplying the usual monthly salary by fourteen and then dividing the total by twelve.

In the case of *blue-collar* employees the contract can again be terminated with or without notice, together with payment of an indemnity, according to the following schedule:

Service	*Notice*	*Indemnity*
Two months to one year	Five days	Five days' pay
One to two years	Eight days	Seven days' pay
Two to five years	Fifteen days	Thirteen days' pay
Five to ten years	Thirty days	Twenty-six days' pay
Ten years and over	Sixty days	Fifty-two days' pay

The indemnity is calculated according to the formula: Daily pay = Usual daily wage × 25 × 14 divided by 25 × 12. Where notice is given, the indemnity is halved.

In all cases the indemnity is a tax-deductible expense for the employer. Although not protected in the event of employer insolvency, employee claims have a high priority. Should the employer decide to pay more than the statutory indemnity, either on dismissal or on retirement, the tax deductibility remains valid. The indemnity is taxable for the employee, and operates via a withholding tax deducted by the employer. However, the rate of tax is substantially lower than on normal earned income, and runs at 10 per cent up to Dr 1 million (£3,250), and thereafter at 15 per cent.

Grounds for dismissal

The reason for the dismissal does not have to be given at the moment of dismissal, but would have to be given were the employee to challenge the termination in the courts. In broad terms, dismissal must merely be 'objective', as opposed to 'abusive'. The notice of dismissal must be in writing and delivered to the employee (or the employee's legal representative) in person.

Most practitioners would argue that dismissal on grounds of poor performance or persistent misconduct should be preceded by adequate warnings to the employee: in the case of poor performance they could take the form of appraisals signed and accepted by the employee.

In companies employing more than seventy employees the law requires that a personnel handbook, setting out company rules and employee rights and obligations, should be agreed with enterprise union representatives (although compliance is not widespread). Breaching agreed rules may provide grounds for dismissal, but this has to be the last resort, and be preceded by: an oral warning; a written warning; deductions from pay not exceeding a quarter of a day's pay; or suspension for a period without pay.

Protected employees

Notice of dismissal may not be given to any employee on leave, to any pregnant women or women on maternity leave, to any trade union official or health representative, depending on the particular office, either during office or for a period afterwards.

Appeal and reinstatement

Appeal is via the civil courts, although the first step by an employee might be to contact the Ministry of Labour to seek mediation. If a dismissal is found to be unfair a court can order reinstatement, should the employee wish. Where this takes place the employee must repay the severance indemnity but is entitled to any remuneration due from the time of dismissal to the court ruling, plus interest at 1·5 per cent per month. However, an employer can immediately re-dismiss an employee, with the appropriate indemnity. In practice such protracted disputes are rare, and in most instances the parties will come to a settlement involving a higher than statutory indemnity payment to allow the contract to be terminated by mutual agreement.

All dismissals must be notified to the employment service, OAED.

Appendix

OAED (the state manpower and job placement agency)
Thrakis 8
166.10 Glyfada
Athens
tel. +30 1 993 2589
fax +30 1 993 7301

Ministry of Labour
40 Piraeus Street
Athens
tel. +30 1 523 2110

Centre for Occupational Health and Safety
6 Dodekanissou Street
17456 Alimos
tel. +30 1 999 9566

General Secretariat for Equality
2 Mousseou Street
Plaka
Athens
tel. +30 1 321 2094

Federation of Greek Industries (represents employers in manufacturing and services, and negotiates general collective agreements with the Confederation of Labour, the trade union centre)
5 Xenofontos Street
105 57 Athens
tel. +30 1 323 7325
The federation also has regional offices.

Greek Personnel Management Association (personnel managers in companies with at least fifty employees with a distinct personnel function may join; the association produces salary and benefit surveys and seeks to exercise influence on employment matters)
3 Karitsi Street
G-10561 Athens
tel. +30 1 322 5704

British–Hellenic Chamber of Commerce
25 Vas. Sofias
10674 Athens
tel. +30 1 721 0361
fax +30 1 721 8571

ETBA
The Greek Investment Bank (ETBA) produces an investment guide to Greece with a labour law chapter. The guide is updated every two or three years, and is written for ETBA by the Greek office of Coopers & Lybrand.

Coopers and Lybrand
Abacus House
9 Semitlou Street
115 28 Athens
tel. +30 1 77 10 134

ETBA
18 Panepistimiou Street
10672 Athens
tel. +30 1 323 7981
fax. +30 1 362 1023

Stedima SA, business consultants
3 Hadjiyanni Mexi Street
11528 Athens
tel. +30 1 723 4771
fax +30 1 721 1414

Acknowledgements

The authors are indebted to Stedima SA Business Consultants.

6

The Irish Republic

The determination of individual terms and conditions of employment is primarily by collective agreement at company level with a floor of statutory provisions which have had an increased impact over the last 25 years, although without undermining the essentially voluntaristic character of the system. Since 1987 pay increases and hours reductions have been the subject of national tripartite agreements set to run until 1993.

The 1937 Irish constitution guarantees all employees freedom of association. Exceptions include members of the defence forces and the police (Garda Síochána). However, trade unions are not obliged, constitutionally or otherwise, to accept every applicant into membership. Employees are also guaranteed a constitutional right to take part in the decision-making processes within the rules of a trade union. There is also a right, confirmed by the courts, not to belong to a trade union.

Contracts of employment

The contract of employment is an enforceable bargain agreed between two people or parties under which the employee agrees to provide services to the employer in return for a wage.

Terms and conditions laid out in contracts must not contravene the constitution in any way. They may incorporate terms and conditions laid down in national or industry-wide collective agreements. There is no obligation to provide a written contract. However, most white-collar and managerial employees have written contracts.

Employees covered by the Minimum Notice and Terms of Employment Act 1973, that is, those who have been employed continuously for thirteen weeks by the same employer and work at least eighteen hours a week, are entitled to a written statement of:

- The starting date of employment.
- Pay details, including overtime, commission and bonus.
- The method of payment (whether weekly or monthly).
- Hours of work and overtime arrangements.
- Holiday entitlements.
- Sick pay arrangements and pension schemes.
- Periods of notice or, if the employee is employed on a fixed-term contract, the date of expiry.

This written statement must be provided by the employer at the employee's request within one month of hiring. It does not necessarily constitute a contract, being merely seen as evidence of the main terms and conditions of employment established between the two parties.

Express terms of a contract are those which have been definitely agreed between the two parties and may or may not be in writing. However, a contract also contains 'implied terms' dealing with the rights and duties of the employer and employee which are laid down in the constitution. On the employer side these include:

- The duty to maintain the employment relationship, although the employer is not under any initial obligation to provide an employee with work.
- The duty to pay agreed wages. If this is not met, the employee is entitled to sue the employer for the amount due.
- The duty to provide a safe working environment. An employer must take care to ensure that equipment, premises and materials do not present any hazards.
- The duty to treat an employee with respect.

Employees are also under certain implied obligations when they enter into an employment relationship, including:

- The duty to obey. Refusal to obey a legitimate instruction from an employer could result in summary dismissal for gross misconduct.
- A duty of fidelity. Employees must not place themselves in such a position that their own interests conflict with their duty to their employer. For example, they must not compete with the employer, divulge confidential information or trade secrets, or knowingly impede the employer's business.
- The duty to treat an employer's property with due care and respect.

Part-timers

Part-time workers represent 8 per cent of the work force in Ireland, the number of employees on part-time contracts having increased by nearly a quarter between 1975 and 1988. There is no statutory definition of part-time work. However, under current legislation, only those employees working more than eighteen hours a week are entitled, on a pro-rata basis, to benefits enjoyed by full-time staff, covering holidays, workplace health and safety, dismissal notices, redundancy payments and redress for unfair dismissal. This topic has been under discussion and a Bill reducing the minimum requirement of eighteen hours to eight hours for employees with thirteen weeks' continuous service becomes law in 1991.

Some company agreements do, however, already provide for pro-rata treatment of part-time workers.

Temporary workers

There is no statutory obligation for the contract of employment of an employee engaged on a temporary basis or for a fixed term to be written and no statutory guarantee of access to benefits enjoyed by permanent staff. Workers employed on a temporary basis are often employed on contracts of less than one year in order to avoid certain statutory obligations connected with length of service, such as the right to claim unfair dismissal after one year's service. There is no statutory definition of 'fixed-term contract' and employers do not have to justify recourse to employing staff on a temporary basis.

Restrictive covenants

Contracts may contain clauses restricting the activities of an employee or former employee in order to prevent unreasonable harm to the employer's interests. However, the restraint must not be unreasonably extensive in area, time or subject matter and it must not be unreasonable in relation to the interests of the employee in the maintenance of the employee's freedom. An employer may lawfully protect trade secrets and confidential information. In the absence of an express restrictive covenant, an employee's implied duty of faithful service will operate, obliging the employee not to solicit a former employer's customers or disclose confidential information and trade secrets during employment and, in some cases, in the period following employment.

Employee inventions

Although there is no stipulation in the Patents Act 1964 that employees' inventions belong to the employer, it is generally held that they do, as a high proportion of inventions are produced by employees in the employer's time and on the employer's premises. The contract of employment will generally set out the position with regard to ownership of inventions. In the event of a dispute, the Controller of Patents and, where appropriate, the High Court will settle the matter, possibly by apportioning the benefit of a patent between the parties.

Individuals' workplace rights

Employee grievances

If an individual has a grievance which cannot be settled through direct negotiation with the employer, the employee may take the matter either to a Rights Commissioner for investigation and recommendation or to the Employment Appeals Tribunal (EAT).

If a claim is taken to a Rights Commissioner the employee or union official must notify in writing both the Rights Commissioner and the employer involved. The hearing between the parties is private and informal. Employees may be represented, if desired, by a union official or friends and may take witnesses along. Employers may be represented by a solicitor or a representative of the FIE (employers' federation). Any party unhappy with the decision of a Rights Commissioner may appeal to the Labour Court, whose decision is then binding.

Employment Appeals Tribunal. If either party objects to a Rights Commissioner hearing the case, a grievance can be taken straight to the Employment Appeals Tribunal. The EAT must be notified in writing and advised of the fact that the case has not been heard by a Rights Commissioner. EAT hearings are quite informal although public; parties may request a private hearing if desired. Parties who are dissatisfied with an EAT decision may appeal to the High Court.

Disputes regarding equal pay may be referred to an Equality Officer for investigation in private and a recommendation. Either party to a dispute may appeal within forty-two days to the Labour Court against an Equality Officer's recommendation (see also below).

Labour Relations Commission. Rights Commissioners and Equality Officers act under the auspices of the Labour Relations Commission, a body set up under the Industrial Relations Act 1990. The LRC will take over many of the advisory and conciliation functions of the Labour Court, although the Labour Court will still function as a court of last resort. The thinking behind this is to promote conciliation and negotiation between parties instead of automatically referring matters to the Labour Court. (Conciliation machinery and the functioning of the Labour Relations Commission will be dealt with in greater detail in volume 3 of the European Management Guides series.)

Data protection

Under the 1987 Data Protection Act employees are entitled to access to automated personal data in relation to them and to have inaccurate data rectified or erased. Data must not be disclosed for any reasons incompatible with the purpose for which they are kept. This covers all personal information on employees kept on computer, such as data on the employee's health, family and social circumstances or political opinions. Persons keeping specified kinds of data are obliged to register with the Data Protection Commissioner, who is responsible for investigating complaints made by data subjects.

Working time

Hours of work

Under the Conditions of Employment Acts 1936 and 1944 the limit to the number of hours that adult industrial workers may work is forty-eight hours in any week and nine hours in any day. For day work, employees may not continue working beyond 20.00 on an ordinary working day or 13.00 on a half-day. These restrictions cover all industrial workers except those employed in agriculture, fishing, domestic work and commerce. However, they do not apply to industrial work in printing and publishing newspapers or in broadcasting and telephone services.

The current minimum working week is eighteen hours if an employee wishes to enjoy, pro rata, the rights and benefits of a full-time employee, although a Bill to reduce this minimum to eight hours will become law in 1991.

Young persons may not start work earlier than 08.00 or finish after 20.00, nor may they work more than eight hours a day or forty hours in any week.

Workers in retail and wholesale shops, hotels, restaurants and cafés are governed by the Shops (Conditions of Employment) Acts 1938 and 1942. The Acts set the maximum working day at eleven hours, the maximum working week at fifty-six hours in hotels and forty-eight hours in shops, and set absolute weekly, monthly and annual limits of sixty-six, 244 and 2,900 hours respectively in the case of hotels.

Actual working time, typically 39 hours in industry, is regulated by collective agreements, at either industry or company level, and have also been subject to the national tripartite agreement, the National Recovery Programme, concluded in 1987. Under the 'Framework Agreement on Working Hours' concluded in January 1989, a product of the commitment to talks on working hours laid down in the tripartite agreement, companies undertook to cut working hours by one hour for employees working forty or more hours a week. Around 70 per cent of 200 agreements surveyed in 1989 provided for a straight one hour off, usually on a Friday afternoon. Additional annual leave was the next most popular method for day workers.

Shift work

The length of a shift must not exceed nine hours. A worker may not work two consecutive shifts and there must be a minimum rest period of eight hours between shifts. Average hours worked may not exceed forty-eight hours per week in any three consecutive weeks. Exceptions may be made to these limits for some continuous process operations, in which case a licence must be obtained from the Minister of Labour.

Overtime

Overtime is allowed under the Conditions of Employment Acts 1936 and 1944 at the rate of either two hours a day, or twelve hours in any week, or thirty-six hours in any four consecutive weeks, or 240 hours in any year. These limits may be exceeded if permission is granted by the Minister of Labour. Any statutory overtime — that is, work in excess of the maximum forty-eight hours a week — must be paid at a minimum of time-and-a-quarter. In practice, agreed overtime supplements are payable for hours worked in excess of the normal working week, currently thirty-five to forty hours, usually at time-and-a-half for the first five hours.

Overtime limits for young persons are two hours a day, ten hours a week, thirty hours over four consecutive weeks or 200 hours a year.

Breaks

For day workers, no employee should work a period exceeding five hours without a thirty-minute rest period. Shift workers must be given a rest period of at least fifteen minutes in each shift, no later than the fourth hour of the shift.

Time off

Annual leave

Under the Holidays (Employees) Act 1973 employees are legally entitled to three working weeks' paid leave a year, with current average actual provision of 20 days. Employees qualify for the full entitlement if they have worked for the same employer for 1,400 hours in a leave year (1 April–31 March) or 1,300 hours if the employee is under 18. Otherwise an employee is entitled to one and a quarter days' paid leave for every full month worked (120 hours).

Public holidays

Employers are not obliged to grant the actual public holiday as time off to the employee. They are free to choose one of the following: a paid day off on the day itself; a paid day off within a month; an extra day's annual leave; the closest Church holiday before or after the public holiday; an extra day's pay.

There are eight public holidays: 1 January, the first Monday in August, St Patrick's Day, the first Monday in June, the last Monday in October, Easter Monday, Christmas Day, St Stephen's Day. Leave records must be kept by the employer for a period of at least three years.

Sick leave

Employees are entitled to paid sick leave in the form of a 'disability benefit' payable by the Department of Social Welfare from the fourth day of absence from work on production of a medical certificate. From the fourteenth day, a pay-related statutory supplement may also be added. Most employees are also covered by contractually agreed sick pay schemes which vary widely from full normal basic salary to a fixed flat-rate payment. According to an employer survey, 84 per cent of surveyed companies in the Dublin region operated some kind of sick pay scheme, the majority offering full pay from the first day of illness. Forty-one per cent of the companies offered benefit for twenty-six weeks or more. In the case of employees covered by a company scheme and entitled to social welfare benefits, it is customary for the employer to deduct the amount of social welfare benefit from the amount of sick pay that would normally be paid.

Maternity leave

Under the Maternity Protection of Employees Act 1981 all female employees employed on a permanent basis for at least eighteen hours per week are entitled to a minimum period of maternity leave of fourteen weeks. The leave must commence at least four weeks before the end of the expected week of confinement and finish no earlier than four weeks after the end of that week. There is no service requirement for this benefit, although to qualify employees must have paid social security contributions for thirty-nine weeks during the preceding twelve months. This leave is paid by a social insurance fund at a rate of 70 per cent of earnings up to a ceiling of Ir £154. Employees may take an additional four weeks' unpaid leave if desired. The 1981 Act also establishes a right to unpaid time off for ante-natal and post-natal care as necessary. The employee's right to return to work is conditional upon written notification of the intention to return four weeks before the due date of return, followed by written confirmation two weeks before the due date of return.

Educational and parental leave

There is no statutory provision for educational or parental leave. There are well defined provisions for career breaks in the civil service. In the private sector, career breaks tend to operate on an *ad hoc* basis. However, some companies in the banking and financial services sector have schemes entitling employees to an unpaid break from a minimum of one year to a maximum of three years. Employees have a right of return to their previous status provided they give two months' notice of their intention to return.

Trade union or civic activities

There is no statutory right to time off for trade union activities. However, in practice, shop stewards would be allowed to carry out their activities during working hours with no loss of pay. Some house agreements contain time-off provisions for trade union duties.

Retirement

The statutory retirement age is 65 years for both men and women. There is no statutory early retirement provision, and an employer survey has shown that the majority of both male and female employees tend to work until the age of 65. Under the state pension scheme, pensioners are entitled to a flat-rate benefit averaging around 25 per cent of the average industrial wage, with supplements for the over-80s. Entitlement to benefit is based on the employee's contribution record.

Company pension schemes to complement the state benefit are widespread. An employer survey found that the vast majority of companies surveyed operated some kind of scheme for their employees, 79 per cent of which required some form of contribution from employees.

Under the Pensions Act 1990 a regulatory body, the Pensions

Board, was created to monitor occupational pension schemes and check their compliance with the law. Pension trustees must report to the board every three and a half years to prove that sufficient assets are held to meet minimum benefit liabilities. The Act also provides for the preservation of benefits when pension scheme members leave employment before normal pensionable age but after a period of five years' service under a scheme.

Equal treatment

The Irish constitution guarantees equality of all citizens before the law. The Employment Equality Act 1977 provides for equal treatment of men and women on matters relating to employment training and promotion, conditions of employment, access to employment and dismissal. This law covers all employees except the defence forces, police and prison service posts and certain limited categories of employee where persons of one sex are required for a particular job. This legislation is concerned solely with equality of treatment between men and women and so does not extend to equality of treatment for disabled employees or ethnic minorities.

Discrimination is defined either as direct, when a person is treated less favourably than another on grounds of sex or marital status, or indirect, such as imposing a non-essential condition upon employment which a substantially greater proportion of people of one sex or category of marital status could comply with.

The Employment Equality Agency, which was created by the 1977 legislation, has the general function of promoting equality of opportunity between men and women in relation to employment and of overseeing the operation of the legislation and the Anti-discrimination (Pay) Act 1974. Claims of discrimination may be taken for advice to the Employment Equality Agency and then referred to the Labour Court within six months of the alleged discrimination. The Labour Court will either settle the dispute through an Industrial Relations Officer or by reference to an Equality Officer, who will investigate the dispute and make a recommendation.

If discrimination is proved, the recommendation will require action to comply with the law, including up to 3 years' pay arrears. If the employee leaves or is dismissed, compensation of up to 104 weeks' pay may be ordered, together with a fine of up to Ir£100 on summary conviction for failure to follow a course of action recommended by the court, plus Ir£10 for each day the offence continues.

If an employee believes that they were dismissed because they brought an action against their employer on grounds of alleged discrimination, or were justified in resigning because of the employer's conduct, or that their dismissal was for discriminatory reasons, they may appeal within six months of the dismissal or resignation to the Labour Court or a court of law. The Labour Court will either:

- Dismiss the claim.
- Order the employer to pay compensation.
- Order the employee's reinstatement or re-engagement in a similar position.

A Court of Law may:

- Impose a fine of up to Ir£100 on summary conviction and up to Ir£1,000 on conviction on indictment.
- Impose further fines (up to Ir£100 plus Ir£10 per day) for refusal to comply.

Work environment

The Health, Safety and Welfare at Work Act 1989 sets out the general duties of employers and employees relating to health and safety at work. Under the Act employers employing more than ten staff are required to identify the hazards and assess the risks at the place of work and draw up a written Safety Statement setting out the arrangements in force to safeguard health and safety and the co-operation required from employees to achieve it. Management should then ensure that working practices conform to the Safety Statement. Areas to be covered by the Safety Statement include:

- Identifying the hazards in the workplace and assessing the risks arising from those hazards.
- Specifying the manner in which safety, health and welfare are to be secured and giving details of arrangements made and resources available for the purpose.

- Specifying the co-operation required from employees in safety and health matters.
- Giving the names and job titles of people responsible for health and safety in the organization.
- Giving details of the arrangements for consultation with employees on health and safety.

The Safety Statement should be brought to the attention of employees at least annually and whenever it is revised. Under the Act an inspector can require revisions to be made to the Safety Statement if it is found to be inadequate. It must also be kept up to date regarding the names of responsible persons, changes in risks, new machinery, new personal protection equipment and new legislation or standards.

The Act also provides for consultation between employers and employees to help ensure co-operation in the prevention of accidents, although it does not set out any specific mechanisms. Employees are entitled to appoint a safety representative to consult employers and represent employees on health and safety matters. This representative is also allowed to carry out inspections at the workplace to identify hazards and risks.

Under the 1980 Factories Act, if more than five persons are employed the employer must provide facilities at the workplace for boiling water and taking meals which must be 'suitable and adequate'.

There is no legislation protecting the rights of non-smokers at the workplace.

Individual termination of contract

Grounds for dismissal

There are two forms of dismissal: dismissal for reasons relating to the employee's conduct and circumstances, and dismissal for reasons relating to the company's circumstances (redundancy).

Grounds for dismissal relating to the employee's conduct range from lack of competence, sickness or injury, absenteeism or lateness, disobeying orders, gross misconduct, general behaviour off the job if it is deemed to be relevant, and criminal offences. Under the Minimum Notice and Terms of Employment Acts 1973 and

1984 a minimum period of notice must be given by employers prior to the termination of a contract of employment (see above). However, if an employee is deemed to be guilty of gross misconduct, they are likely to be given summary dismissal. The kind of behaviour likely to be seen as gross misconduct is usually set out in house agreements and varies according to the industry and workplace.

Redundancy is covered by the Redundancy Payments Acts 1967–84 and is defined as the termination of employment where an employer needs fewer employees or fewer employees of a particular kind. This can arise for several reasons, such as reorganization, a trade recession or the closure or partial closure of the business. The procedures in each case are set out below.

Procedure

Non-economic dismissal. There is no statutory procedure to be followed by employers wishing to dismiss an employee for factors related to their person or conduct. However, the following guidelines have arisen from case law, custom and practice:

- The employee must be duly warned.
- The employer must point out the consequences of the employee's conduct.
- The employer must investigate the matter fully, acquaint the employee with the facts and allow them to reply to the allegations and offer a defence.
- Dismissal procedures must follow those outlined in a written statement which each new employee receives from the employer and which is usually combined with the written statement of general terms and conditions of employment (see above).

If requested by the employee, an employer must, within fourteen days of the dismissal, give written particulars of the grounds for dismissal. There are no penalties in the Act for contravention of this provision, but the employer's compliance will be taken into account in determining the amount of compensation awarded if a dismissal is found to be unfair.

Redundancy procedures and payments. If the dismissal is on economic grounds the employer must give the employee written notice at least two weeks before the date on which the dismissal is due to take effect. A copy of the written notice must be sent to the Department of Labour. A longer period of notice may be required by the Minimum Notice and Terms of Employment Acts. Lump sum redundancy payments must be given to employees not later than the date when the redundancy takes effect. Minimum redundancy payments under the Acts are as follows:

- One week's pay, plus
- Half a week's pay for each year of continuous employment between the ages of 16 and 41 years, plus
- One week's pay for each year of continuous employment over the age of 41 years.

These payments cover weekly earnings up to a ceiling of Ir£250.

Employers usually qualify for a rebate of 60 per cent of the statutory lump sum payment from the Redundancy Fund, which may be reduced to 40 per cent at the discretion of the Minister for Labour if the employer does not comply with notice requirements.

The employer must give the employee the lump sum payment on the date of dismissal, together with form RP2, which is the certificate of redundancy. When the lump sum has been paid, the employer can seek a rebate from the Redundancy Fund on form RP3. Both these forms can be obtained from the Department of Labour.

Period of notice

Under the Minimum Notice and Terms of Employment Act 1973, employees who have been in continuous service with the same employer for thirteen weeks or more are entitled to a minimum period of notice on dismissal as follows:

- Thirteen weeks to two years' service: one week's notice.
- Two to five years' service: two weeks' notice.
- Five to ten years' service: four weeks' notice.
- Ten to fifteen years' service: six weeks' notice.
- Fifteen years' service or more: eight weeks' notice.

Individual contracts may require a longer period of notice which is legally binding unless there is mutual agreement allowing for alteration of the terms of the contract. The right to notice may be waived if payment is offered in lieu.

Fair and unfair dismissal

Under the Unfair Dismissals Act 1977 the onus is upon the employer to prove that a dismissal was not unfair. A dismissal is deemed unfair if it is caused by:

- An employee's membership of, or activity on behalf of, a trade union.
- An employee's religious or political opinions.
- The race or colour of an employee. Civil or criminal proceedings, actual or proposed, against the employer, to which an employee is a party or a witness.
- The exercise by an employee of her rights under the Maternity Protection of Employees Act 1981.
- Redundancy, if the selection of an employee for redundancy is deemed to be unfair.
- An employee's participation in a strike or other industrial action (if one or more other participants were not dismissed, or were subsequently offered reinstatement or re-engagement).
- Lock-out, if an employee is not offered reinstatement upon the resumption of work.

A dismissal is deemed not unfair if the employer can prove that it arose from:

- The conduct of the employee.
- The ability, competence or qualifications of the employee.
- Any regulation or statute which prevents the employer from continuing to employ the employee.
- Redundancy, if the appropriate selection procedures are fully observed.
- The expiry of a fixed-term or specified-purpose contract, provided the contract is written, is signed by both parties, and specifically states that dismissal at the end of the contract is not subject to the Act.

Recourse to law, and settlements

An employee who wishes to lodge a claim of unfair dismissal must notify the Rights Commissioner within six months of the dismissal, and the commissioner will then make a recommendation.

The claim goes to the Employment Appeals Tribunal if:

- The employee wishes to take the case direct to the tribunal.
- The employer objects to the hearing by a Rights Commissioner.
- The employer fails to comply with a commissioner's recommendation.
- If either party appeals within six weeks against a commissioner's recommendation.

The case will then go to the Circuit Court if:

- Either party appeals against a determination of the tribunal within six weeks.
- The Minister for Labour decides to proceed against an employer for non-compliance with a determination of the tribunal.

If unfair dismissal is proved, one of the following courses of action may be recommended:

- Reinstatement with arrears of pay from the date of dismissal.
- Re-engagement in the same or a similar position on conditions specified by the Rights Commissioner, tribunal or court.
- Compensation by the employer, up to a maximum of 104 weeks' pay. Most contested dismissals end in payment of compensation.

Final formalities

Employees are normally under obligation to work out any period of notice given. However, if the employee wishes, and the employer is in agreement, the employee may leave before the notice expires. The employee must notify the employer in writing. Under the Redundancy Payments Act 1979 employees are allowed 'reasonable' paid time off during the final two weeks of notice to seek new employment.

Employers must honour any obligations concerning the payment of wages, holiday money and sick pay. Employers are under no obligation to give employees work to do during their period of notice.

Appendix

Department of Labour
Mespil Road
Dublin 4
tel. (01) 765861

Training and Employment
Authority (FAS)
27 Upper Baggot Street
Dublin 4
tel. (01) 685777
fax (01) 682691

Federation of Irish Employers
(FIE)
Baggot Bridge House
84–6 Lower Baggot Street
Dublin 2
tel. (01) 601011
fax (01) 601717

Irish Congress of Trade Unions
(ICTU)
19 Raglan Road
Dublin 4
tel. (01) 680641
fax (01) 609027

The Association of Management
Consulting Organizations
(AMCO)
Confederation House
Kildare Street
Dublin 2
tel. (01) 779801

Institute of Personnel
Management
35–9 Shelbourne Road
Ballsbridge
Dublin 4
tel. (01) 686244

7

Italy

Terms and conditions of employment in the private sector are regulated by a pattern of collective agreements which are binding on the whole of the sector which they cover. Employment contracts are established within the broad legal framework underpinned by the Civil Code. Collective and individual rights on issues such as termination are both rooted in constitutional principles and the Workers' Statute, passed in 1970 in the wake of the upsurge in industrial militancy in the late 1960s. Termination procedures are complex, and a recent major amendment to the legislation in this field extending protection against unfair dismissal to all employees has done little to rationalize a piecemeal system. Conflicting laws in this area can still make the outcome of a case unpredictable, a situation compounded by the possibility of uncertainties in the interpretation of the law.

Economic relations occupy a central part of the Italian constitution and a corresponding emphasis is put on the protection of employee rights. Article 35, for example, states that 'The Republic shall protect labour of all types and however used,' while article 41 affirms that, although 'private enterprise' shall be unrestricted, it must not conflict with 'the good of society, the needs of safety, liberty or human dignity'.

The constitution provides all employees with the basic rights: to work, to be paid for that work, to enjoy a weekly rest day and annual holidays, to receive social welfare and benefits; and the right to freedom of association. Within this broad framework a body of protective legislation and legally enforceable collective agreements has evolved, covering wide sections of industry and many categories of employee, including managers.

Contracts of employment

Contracts of employment for manual and white-collar workers are regulated mainly by the Civil Code, as amended by other specific employment legislation. The most important laws are Act 230 (18-4-62) on fixed-term contracts of employment; Act 604 (15-7-66), as amended by law 108/1990, on individual dismissals; and Act 300 (20-5-70), the Workers' Statute, which guarantees trade union freedoms, gives enhanced protection against unfair dismissal and regulates some areas of recruitment.

Contracts of employment are dependent upon the will of the parties entering into the contract. The form in which contracts are drawn up is, in general, unrestricted, with some exceptions where a written contract is required. Individual contracts must conform with any collective agreements that set out binding minimum pay levels and other conditions.

Employment contracts are personal contracts (*nominato*) and the functions agreed in the contract must be carried out by the employee him- or herself. No substitution is permissible. For a contract of employment to be valid the employee must have 'the capacity and the power to fulfil their functions' and also have the 'legal capacity' to agree to the contract. Although an individual is entitled to become a worker at the age of 15, no one may enter into a contract of employment in their own right without the consent of a parent or guardian until the age of 18.

Form of contract

Generally there is no legal requirement for contracts to be in writing, except in specific cases. It is sufficient for the two parties — the provider of employment (*datore di lavoro*) and the giver of labour (*lavoratore*) — to have made a verbal declaration of willingness for the agreement to be binding. In practice, however, so many collective agreements and agreements for particular categories of employee stipulate that the contract of employment must be in writing that verbal agreements really apply only to blue-collar workers in industry. And even in that case many companies prefer to draw up a written contract for this category.

The following types of contract must by law be in writing:

- Fixed-term contracts.
- Part-time contracts.
- Contracts for employees in sport.
- Training and work experience contracts.
- Contracts for seamen.

In addition, particular types of contractual clauses must be in writing: these include clauses specifying the length of probationary periods and restrictive covenants following termination of contract.

Types of contract

By law employment contracts which are not permanent and full-time are considered 'special'. 'Special' employment contracts are regulated by the appropriate legislation and are permissible either as part of a job creation scheme or because they are inherent in the nature of the work involved — such as sporting activities or the entertainment business. Other 'special' employment contracts apply to, say, domestic work or the work undertaken by building caretakers. The main types, together with their requirements, are as follows:

- *Apprenticeships.* Applicants must appear on a special list at state job placement offices. There are a limited number of apprenticeships for each company, and authorization from the labour inspectorate is required. Apprentices may not be younger than 15 or older than 20, and the maximum duration of the contract is 5 years. Piecework is not permitted.
- *Part-time contracts.* These must be in writing. Collective agreements can establish the percentage of the work force on part-time contracts, and the permitted range of tasks. Overtime working is not permitted, and pay and benefits must be pro rata with full-time employees'.
- *Solidarity contracts*, intended for periods of business difficulty, are divided into two types. *Temporary* solidarity contracts may be used where economic circumstances require a temporary reduction in working hours with a proportionate reduction in pay. The difference in pay is then partially supplemented by the Cassa Integrazione Guadagni (CIG) wage-support fund for up to two years. (The CIG is a form of long-term lay-off fund, jointly financed by employer contributions and state funds, which effectively forms the basis of income support during periods of unemployment: its functioning will be dealt with in detail in volume 3 of the European Management Guides series dealing with industrial relations.) *Permanent* solidarity contracts can be used as part of a job creation scheme where employees work shorter hours for a proportionately lower permanent level

of pay. Under these circumstances individuals (usually young people) can be recruited 'by name' (that is, they do not have to be taken from the unemployment lists) on permanent contracts. Employers benefit from reduced social security contributions.

- *Training and work experience contracts.* Applicants can be recruited 'by name' and should be aged between 15 and 29 years. The maximum duration of the contract is twenty-four months. The employer benefits from economic and procedural exemptions, such as reduced social charges and free choice in recruitment.
- *Fixed-term contracts.* These are permitted in certain circumstances, such as when necessitated by the inherent nature of the work involved, when cover is required for an individual on leave (e.g. maternity, military service) or to accomplish a specific task. The contract must be in writing and a copy sent to the employee. It is renewable only once and for a term no longer than the original. Pay and benefits should be pro rata with full-time employees'.
- *Contracts for managers* (*dirigenti*). The special nature of this type of contract embraces enhanced powers and responsibilities within and on behalf of the enterprise; rights of supervision over other employees; exclusion from social legislation such as the duty to sign on at a placement office and to carry an employment record (*libretto di lavoro*); exclusion from legislation on working hours and on fixed-term working. However, *dirigenti* are covered by their own collective agreements setting minimum pay levels.

Rights and obligations

Article 2094 of the Civil Code defines a 'subordinate employee' (*prestatore di lavoro subordinato*) as one who 'undertakes, for remuneration, to collaborate within the enterprise, by providing his own labour, intellectual or manual, as an employee under the direction of the employer'.

The concepts of 'dependence' and 'subordination' lie at the core of the employment relation in Italy and confer on the employer both a hierarchical responsibility for, and authority over, the employee. The employer is 'the head of the enterprise and on him depend hierarchically his collaborators' (article 2086, Civil Code). The employer has the right to manage and direct the enterprise and the power to discipline its employees; the employee has a reciprocal duty of obedience, within the limits of his or her assigned function.

Right of transfer

If the question of transfer is not covered by company rules or collective agreement, and no specific limitations have been imposed by the agreement of the parties, the employer is generally held to have the right to transfer employees within reasonable bounds according to their function and the locations in question. However, trade union officials and employee representatives on works committees (*consigli di fabbrica*) may not be transferred for one year following the cessation of their duties without the written authorization of the relevant union at district level.

Loyalty, obedience and diligence

In undertaking to 'collaborate within the enterprise' (Civil Code, article 2094) the 'subordinate employee' makes a commitment to dedicate him- or herself, personally, to the fulfilment of the functions agreed in the contract, but only within the limits of those functions. The nature of the contract of employment, then, as far as the employee is concerned, is expressed in the dual obligations of loyalty and obedience. These duties of loyalty and obedience also entail a duty of diligence, which, according to article 2104 of the Civil Code, 'conforms with the nature of the work being undertaken, with the interests of the company and with the interests of national production'.

The employer's duty of care

Apart from their duty to pay the employees, employers also have a duty to safeguard employees' physical and psychological well-being. It is up to the employer to ensure that the employee has the necessary facilities to fulfil the duty of diligence accepted when taking on paid employment.

Grading

Article 96 of the Civil Code states that employers should inform employees when they start work which grade and category of employment their job falls into.

Tasks

Employees can be required to undertake only tasks which fall within their own job grade or, on a temporary basis, within the grade above or equivalent tasks without loss of pay.

Probationary period

In general an employee's probationary period must be specified in writing. (One exception would be where an enterprise has a clear practice set out in its personnel handbook.) Both parties can terminate the contract of employment during this period without giving notice and without payment, except for services already rendered by the employee. Should the probationary period be completed satisfactorily it is added to the employee's length of service once the contract has been made permanent. Probationary periods can last between twelve days and six months, depending on job grade.

Restrictive covenants (*patto di non concorrenza*)

Article 2125 of the Civil Code rules that a restrictive covenant will be null and void unless certain conditions are met:

- The agreement must be in writing.
- It must contain compensatory provision for the employee.
- It must specify limits on the purpose, period and scope of the agreement.
- It cannot be valid for more than five years in the case of executives or for more than three years in the case of other employees.

Contractual requirements

In *industrial* companies a verbal agreement is generally sufficient for blue-collar employees (*operai*), except for fixed-term contracts or if a collective agreement requires a written contract. For white-collar staff (*impiegati*) the letter of engagement should include the date of recruitment, the grade, the starting salary and the length of any probationary period. Contracts for sales representatives and commercial travellers (*viaggatori e piazzisti*) should, in addition, specify the scope and limits of their functions, the sales area for which they are responsible, the minimum period in the year during which the company undertakes to keep them on the road, and any duties which may be assigned to them when off the road. For managers, upon engagement or promotion a written statement must be provided which gives details of duties, remuneration and any terms and conditions more favourable than those contained in the relevant national collective agreement.

In *commercial* companies, managers must be given a written statement on engagement or promotion giving the date of engagement, the duration of the probationary period, the initial place of work, recognition of the current national industry agreement and any possible future changes to the agreement, details of duties and remuneration and the possibility of transfer. For all other employees the contract must be in writing and register the date of recruitment, the duration of the probationary period, grade and remuneration.

Individuals' workplace rights

Individual workplace rights such as rights with regard to information, grievances and discipline are dealt with by collective agreement.

The metalworking industry agreement, for example, states that individual grievances or disputes will be dealt with initially by management and workplace representatives. Failure to agree means that matters will be referred to their respective organizations. In addition the agreement lays down a basic framework for conduct within the organization which includes undertakings such as:

- Mutual courtesy and correctness.
- Superiors to encourage co-operative relations between themselves and subordinates.
- The organization to ensure that no employee is placed in an equivocal position regarding individuals, including their immediate superiors, whom an employee might need to refer to for help or instructions.

Working time

Hours of work

The statutory maximum working day is eight hours, with a maximum working week of forty-eight hours. These maxima were established by decree 692 (15 March 1923) and became statute law with law 473 (17 April 1925). Italians work a nominal six-day week — only Sunday is legally recognized as a rest day — but all these laws have been superseded by collective agreement. In 1990 average working time in manufacturing stood at forty hours a week, or 1,737 on an annual basis. Below we give two sectoral examples.

In the *metalworking* industry, negotiated cuts in working hours have mainly taken the form of additional individual time off rather than a shortened working week. Under the terms of the agreement signed in January 1987 the basic week remained at forty hours, but with sixteen hours taken off the working year in 1987, with a further sixteen-hour reduction agreed in 1991.

In the private *chemical* industry the most recent agreement, concluded in July 1990, fixed the contractual working week at forty hours. However, since 1983 significant cuts have been made in annual working hours, again normally taken as additional individual time off. To date, annual working time has been cut by 68 hours for day workers and non-continuous shift workers, with a seventy-six-hour reduction for continuous shift workers. To cater for flexible working time arrangements required for shift working, the normal working week (forty hours for shift workers) may, with the consent of the works council, be arrived at by averaging out the total number of hours worked weekly over a given period. The period over which the average may be calculated is agreed at company or plant level. However, any extra hours worked as part of these flexible arrangements are paid at a supplementary rate.

The 1990 agreement also fixed the normal working week for day workers at thirty-nine hours, worked over five days. However, in exchange employees must relinquish some of the cuts in working hours previously achieved. The number of hours to be forfeited is still the subject of negotiation and will be decided at company level. The likely outcome is between forty-eight and fifty-two hours.

Overtime and flexible working

Two laws, both of them largely superseded by collective agreement, regulate overtime.

Article 5 of law 473 (17 April 1925) states that, with the agreement of the parties concerned, overtime may be worked but must not exceed two hours a day or twelve a week, or an average equivalent spread over a predetermined period. Overtime is defined as hours beyond the statutory maximum of forty-eight a week.

Law 1079 (30 October 1955) provides that overtime in industrial enterprises may be worked only in exceptional circumstances and on a temporary basis. In all cases of overtime working the employer must pay overtime rates to the workers involved plus an extra sum, equal to 15 per cent of the overtime rate paid, into the unemployment fund (*fondo per la disoccupazione*). In addition the regional labour inspectorate must be informed twenty-four hours before the overtime is due to be worked. The inspectorate may forbid or restrict the overtime, and a fine not exceeding L 200 (£0.10!) a day per worker involved may be imposed in cases of infringement (and doubled in cases of subsequent infringement).

In practice, overtime — work beyond agreed hours in this case — in most sectors is regulated by collective agreements, as outlined in the examples below.

In *metalworking* overtime, which must be used only in exceptional circumstances, is limited to two hours a day and eight hours a week. Under the three-year national agreement signed in January 1987 the maximum annual limits are: 150 hours for all workers; 200 hours for workers in companies with fewer than 200 employees; 200 hours for workers in companies carrying out ship and aircraft repairs, and/or marine launching and testing, irrespective of company size; 210 hours for workers in companies carrying out maintenance, installation and assembly work, again irrespective of company size. Employers must inform works councils of any planned overtime working.

However, the maximum number of hours' overtime permitted each year from a worker without informing the works council remain at thirty-two hours a year for shift workers; forty hours a year for non-shift workers in companies with fewer than 200 employees; and 32 hours a year for those in companies with more than 200 employees.

Within the limitations mentioned, overtime, night work and working on public holidays are obligatory if required by the employer unless there is a justified motive for preventing it, and provided the appropriate authorities are informed.

The agreement signed in 1990 in the private *chemical* industry states that overtime and supplementary working may take place only in exceptional circumstances, should be of a temporary nature and should not lead to a reduction in the size of the work force. Supplementary working is defined as time worked between contractual hours and the maximum limit established by law. Overtime is defined as time worked in excess of the maximum legal limit. Overtime must be agreed with the works council in advance. Employees working overtime will be paid at premium rates and must be given time off in lieu. The company is obliged to provide monthly information to the works council on the number of overtime and supplementary hours worked (broken down by service or department) as well as the reason for having recourse to overtime and supplementary working.

However, to provide for flexibility, the normal working week of forty hours may be defined — with the consent of the works council — as an average over a number of weeks to be agreed at company or plant level. If, by the end of the year, a worker has worked more than the maximum overtime limit, he or she will be paid a bonus of 30 per cent for each additional hour up to the maximum of 120 hours per year. As with metalworking, overtime, night work and working on public holidays are obligatory if required by the employer, unless there is good reason and provided the appropriate authorities have been informed.

Time off

Annual leave

Article 36 of the constitution guarantees workers the right to paid annual holidays (*ferie annuali*), which they may not be forced to give up (though no penalties are laid down for its infringement). If, in exceptional circumstances, employees have to work during their leave, they are entitled to double time. Entitlement to leave is acquired following the successful completion of a probationary period. If an employee falls ill during leave the holiday entitlement may be suspended for the period of illness. There are no general legal provisions relating to the length of annual leave, which is determined by collective agreement. Agreements may link holiday entitlement with job grade — and, more often, to length of service.

Most basic leave entitlement is now between four and six weeks a year. The bulk is almost always taken during the month of August, when most factories in manufacturing, for example, shut down. Industry-level agreements mostly provide for a basic annual leave entitlement of four weeks, but an additional week is often given following the abolition in 1977 of five state and religious holidays. The national agreement for the chemical industry, for example, foresees that these five days will be recovered, but detailed provisions are left to negotiation at company level.

In *metalworking*, where workers have a basic annual entitlement to four weeks, this period is increased by one day in the case of employees with between ten and eighteen years' service, and by six days (calculated on the basis of a six-day working week) for those with nineteen years' service and above. The number of weeks taken off together may not exceed three, unless provided for in a company-level agreement. At Fiat, where the basic entitlement is four weeks, a worker with over ten years' service receives one extra day's holiday, with six extra days after nineteen years' service. In addition, all employees at Fiat are entitled to a further five days, to be taken individually; this entitlement is not, however, written into the national agreement, as it followed the abolition in 1977 of a number of public holidays. Moreover, any Christmas shut-down is not taken out of employees' holiday entitlement.

Public holidays

There are ten public holidays (*giorni festivi*). In addition to the national holidays local saint's days are also observed as public holidays in certain cities (such as 24 June in Turin and 7 December in Milan). Law 54 of 5 March 1977 abolished five religious public holidays from 1978, and moved Republic Day and National Unity Day from 2 June and 4 November respectively to the first Sunday in June and November respectively. Many collective agreements subsequently transferred some or all of the seven 'lost' days to employees' annual holiday entitlement, or reduced working time.

Employers are obliged to pay employees normal pay over a public holiday, that is, what they would have earned had they been at work. Article 2 of law 90 (31 March 1954) also lays down that the employer must pay the employee even if the employee would otherwise have been absent from work for whatever reason — illness, injury, maternity leave, annual holiday, and so on. In cases where an employee is at work over a public holiday (*lavoro festivo*) — for example, on shift work or some other agreed exception — collective agreements lay down special bonuses, in addition to a day off in lieu of the public holiday.

Maternity leave

Under law 860/1950 working mothers are afforded statutory protection: this was reinforced by law 1204 (30 December 1971). The mother is forbidden to work during the statutory period of maternity leave, which runs from two months before the expected date of birth to three months after the actual date of birth. In certain jobs — laid down by ministerial decree — this period begins three months before the expected date of birth.

The female employee is entitled to maternity benefit during statutory maternity leave, equal to 80 per cent of her normal pay. The cost of the benefit is borne by the statutory social security organization. Maternity benefits are payable to the employee with the first month's salary or wage following resumption of work. However, the employer must pay an amount in advance which is normally laid down by collective agreement, but which in any case may not be less than half her previous monthly pay.

The law also forbids women from performing dangerous, unhealthy or heavy work from the beginning of pregnancy until the seventh month after the birth. During this period a woman normally assigned to such work has to be assigned to other duties without loss of pay.

Time off for breast-feeding

During the first year of a baby's life the employer must allow the mother two special breaks during the day (one break if working hours are fewer than six) to breast-feed her baby. These breaks must be of one hour's duration each (half an hour if the mother has the use of a room for breast-feeding, or the use of a crèche provided on the premises by the employer). Breaks are counted as normal working time and paid.

Optional parental leave

In addition to maternity leave, a woman may choose to take a further six months' leave during the first year of the baby's life. Her job is protected until the baby is one year old. She is entitled to payments from the sickness benefit fund equivalent to about 30 per cent of her normal salary. Payment is made in advance by the employer, who recovers the amount by offsetting it against employers' social security contributions.

A woman is also entitled by law to time off for nursing a sick child if the child is under three years old. She must present a medical certificate to her employer. This leave is normally unpaid.

The right to special leave can also be exercised by the father when he replaces the mother or has primary responsibility for the child. In the latter case the father must present the mother's employer with a signed statement from her forgoing her own entitlement to the right.

The periods of optional absence from work are counted as periods of employment for the purpose of calculating length of service.

Educational leave

The Workers' Statute of 1970 gave all employees, in companies employing more than fifteen people, certain rights to facilitate study. Article 10 of this law distinguishes between employees attending university (where attendance at lectures is not compulsory) and workers enrolled and regularly attending classes at primary, secondary or vocational schools which are public, state-approved or authorized to award legally recognized qualifications. The latter — but not the former — are entitled to hours of work arranged to facilitate attendance at classes and preparation for examinations. In addition, they may not be compelled to work overtime or during weekly rest periods.

All employees, including university students, are entitled to paid leave in order to sit examinations.

Many collective agreements contain additional provisions on study leave. The pace-setting agreement in this field was the national engineering agreement of 1973, which specified that workers wishing to attend certain schools (that is, state schools or others agreed by employers and unions) were entitled to total paid leave of up to 150 hours over a period of three years. (This leave could be used up in one year by any individual employee.) This allowance was subsequently increased to 250 hours for workers attending primary and middle-school courses (that is, completing their basic education under the Italian system). There are certain limitations on the use of study leave under the engineering agreement. The total cost to the employer must not exceed 0·5 per cent of the payroll; student workers must produce a school certificate certifying their attendance for a number of hours two-thirds more than the number paid for by the employer; no more than 2 per cent of the work force can be simultaneously absent from work for study reasons; and absence from work for study reasons must not hamper normal production. In addition the agreement provides for 120 hours of unpaid leave per annum for all workers, provided that this can be scheduled to take account of the company's organizational and production needs.

Similar agreements exist in most sectors, although the number of hours' leave and the limitations may vary slightly from sector to sector. Sector-level agreements along these lines are aimed at improving the educational standard of workers in general. For individual workers following courses other than the approved ones, other provisions may exist at company level or may be worked out on an informal basis.

Trade union activities

Under section 3 of the Workers' Statute all employees are entitled to ten paid hours' time off per year to attend trade union meetings at the workplace. Trade union delegates are entitled, in addition, to paid leave of absence for the performance of their duties. Entitlement is usually at least eight hours per month. The number of delegates entitled to leave of absence depends on company size (see volume 3 of the European Management Guides series, on industrial relations).

Trade union officials are also entitled to unpaid leave of up to eight days a year for the pursuit of union activities. Collective agreements may improve on the law's provisions.

Employees elected to trade union office at national or provincial level are entitled to unpaid leave of absence at their request at any time during, or for the entire duration of, their term of office.

Family reasons

Most industry-level collective agreements provide for employees to take paid leave if they get married. Entitlement generally ranges from ten to thirty days, starting on the day of the marriage. All white-collar employees are entitled by law to at least fifteen days.

Many collective agreements provide for short leave to be given on the death of a close relative, for attending a funeral or for other family reasons. A typical allowance would be up to twenty-four working hours, or three working days, per year. Many companies may also grant short leave for personal reasons on an informal basis.

Military service

Under law 303 (13 September 1946) employees called to perform compulsory military service are entitled to have their job protected for the duration of the service — that is, to unpaid leave. On completion of military service the employee must present himself for work within thirty days, otherwise the employment contract may be terminated automatically.

In the case of recall to the reserve for any reason, an employee is entitled to paid leave. The amount of pay or benefit is laid down by special laws or may be determined by collective agreement. For white-collar employees, there is a special military benefit fund administered by INPS, the National Social Insurance Institute. At the end of the period of recall the employee must make himself available to his employer within five or fifteen days, depending on whether the recall has been for a period of one month, between one and six months, or over six months. The employee cannot be dismissed during the three months following the resumption of his job, except in the case of justified summary dismissal.

Leave for public duties

The Workers' Statute states that an employee elected as a member of the national parliament, a regional parliament or to other public office is entitled to unpaid leave at any time during, or for the entire duration of, the term of office. The period of elected office must, on the employee's request, be taken into account for the calculation of pension rights. Employees elected to public office who do not wish to be granted unpaid leave are entitled, at their request, to leave of absence with full pay for the time strictly necessary to carry out their official functions.

An employee who is elected mayor or deputy mayor, chairman or vice-chairman of the provincial council is likewise entitled to unpaid leave not exceeding thirty hours a month.

Employees who are carrying out election duties are entitled to three days' paid leave over and above normal leave entitlement.

Employees who voluntarily give at least 250 g of their blood as a blood donor are entitled to paid leave of absence from work for the purpose, and also for the rest of that working day. The employer can obtain a refund of their pay from INPS, the National Social Insurance Institute. The employee must produce a certificate stating that he has taken advantage of the day off to rest, and also a medical certificate from the doctor who took the blood.

Retirement

The official retirement age is 60 for males and 55 for females. Women can opt to work until the age of 60 provided they inform their employer three months before the date on which they would qualify for an old age pension.

The state social security system provides comparatively generous old age pensions, and private company pension schemes are relatively rare. Covered earnings are high, and in 1988 the upper earnings limit under the state scheme was abolished in both industry and commerce, leaving little scope for private supplementary pensions for managers. INPS (Istituto Nazionale della Previdenza Sociale) is the basic scheme which provides old age, disability and survivors' pensions for blue and white-collar workers in industry and commerce and for executives in commerce. Contributions to this scheme must be made for virtually all employees. INPDAI (Istituto Nazionale per la Previdenza dei Dirigenti di Aziende Industriali) provides old age, disability and survivors' pensions for executives in industry.

To qualify for an INPS pension employees must be retired and aged 60 (if male) or 55 (if female) and have paid fifteen years of contributions; or they can be any age provided they have paid thirty-five years of contributions and have retired. To qualify for an INPDAI pension, industrial executives must be aged 65 (males) or 60 (females), with fifteen years' contributions. This arrangement normally replaces INPS. Executives in industry also benefit from a supplementary pension scheme called PREVINDAI.

Early retirement under INPS may be taken with no reduction in benefit provided the employee has thirty-five years of contributions. Executives can retire at any time provided they have made thirty years' contributions: benefits are reduced by 1 per cent for each year of early retirement before the usual retirement age. There is an incentive scheme, usually for use where job losses are required, under which the government may sanction retirement five years before normal retirement age where the employee has made fifteen years' contributions.

Equal treatment

Women

Article 37 of the Italian constitution grants female workers equal rights and equal pay with men for the same work. It also states that women's working conditions must allow the fulfilment of their essential family functions, and guarantees special safeguards for working mothers. While such measures aim to ensure that women's employment and levels of pay are continued after marriage and childbirth, the fundamental assumption behind the law was that childcare and its related family activities were primarily the female's responsibility.

Law 903/1977 prohibits discriminatory acts in employment by granting women the right to equal pay for work of equal value, by allowing women the right to refuse retirement at 55 and by granting the right to continue working until the same age as laid down for men (60).

One innovation under this law was the granting of rights to paid leave and time off to working fathers after the birth of a child, instead of the mother or where she becomes gravely ill or dies.

Effective implementation of the law remains a problem, compounded by indirect discrimination and the lower qualifications of women. Social attitudes also remain a problem, despite formal commitments to equal opportunities. Since 1979 — when many industry agreements provided for greater disclosure of information — consultation and report-back on equal opportunities have been the responsibility of the works council (*consiglio di fabbrica*). The national agreements in the three main industries — engineering, chemicals and textiles — provide for joint commissions to examine possible positive action on equal opportunities in response to the 1984 EC recommendation.

The Minister of Labour is responsible for women's rights and produces an annual report on the functioning of the anti-discrimination laws. Because the courts are so slow in dealing with individual cases of discrimination the trade unions have opposed the setting up of an Equal Opportunities Commission, which, they believe, would merely become another talking-shop.

The disabled

The legislation on recruitment procedures, where the state has the sole right to place individuals in jobs, is aimed, among other things, at providing right of access to employment for the disabled. Under the legislation all organizations, public or private, who employ more than thirty-five people are obliged to recruit a work force in which at least 15 per cent of employees have disabilities which come under the following definitions:

- Invalids (war, service, work, civil), orphans and widows of war invalids.
- Hearing and speech-impaired.

Special provisions also exist for ex-tuberculosis sufferers and the blind.

Enforcement of the law is patchy, and, although employers can buy themselves out of their responsibility, they are rarely called to account.

Ethnic minorities

Law 943/1986 is aimed at regulating the recruitment and treatment of workers from outside the EC, and controlling the flow of illegal immigrants into Italy. The law states that all non-EC community workers resident in Italy should enjoy rights and treatment equal to those of other workers. It also makes provision for a *consulta* (council) to be set up under the auspices of the Ministry of Labour which would deal with the problems faced by non-EC nationals and their families.

Work environment

Basic employer obligations and employee rights on health and safety are outlined in several laws. Article 2087 of the Civil Code requires the employer to adopt the necessary measures for safeguarding an employee's physical and moral well-being, while article 41 of the constitution guarantees safety, freedom and human dignity in the working environment. Article 9 of the Workers' Statute safeguards the health and personal safety of all employees and introduces the notion that employees have a collective right to verify how measures to prevent accidents and work-related diseases are implemented. Detailed regulations on specific industries or processes are covered by presidential decrees.

Local health authorities are entrusted with responsibility for monitoring health and safety matters at work. The overarching purpose of health and safety legislation is to provide:

- A safe working environment ensuring a certain amount of space, light and cleanliness as well as the required temperature, amount of water and safe flooring. Basements or sub-basements may not be designated working spaces, and workers must be protected from accidental contact with machinery. In most industrial companies and in some commercial companies the law states that rooms other than workshops should be provided for changing rooms, bathrooms, dining rooms and toilets. In companies employing more than thirty women below the age of 50 nurseries or feeding rooms must be made available.
- Individual protection for workers. Workers must not wear clothing which could be dangerous. Headgear must often be worn for protection and by employees in certain industries such as food. Other protective clothing should be worn when work processes require it. There are special regulations ensuring the health and safety of women and minors. Pregnant women must stop work for two months prior to confinement and for three months afterwards.
- Materials and processes. Appropriate accident prevention systems should be installed where dangerous substances or processes are in use.

Individual termination of contract

Grounds for termination

An individual contract may be terminated under the following circumstances:

- When the contract expires (in the case of fixed-term contracts) or when the task for which the contract was drawn up is complete.
- If the company ceases trading completely. (This does not include cases where the employer merely changes activity.)
- By mutual consent.
- If it becomes impossible to fulfil the contract because of acts of God or the total incapacity of the employee.
- Withdrawal of one party without the consent of the other.
- If one of the parties fails to fulfil the contract.

If withdrawal from the contract is on the part of the employee through resignation, then there are no legal restrictions, although most collective agreements stipulate periods of notice (see below). If the withdrawal is on the part of the employer through dismissal, then there are significant limitations. They can be divided into two categories:

- Instances where dismissal is not permitted except for a very serious reason — 'just cause' (*giusta causa*).
- Instances where a legitimate reason is required — 'justified motive' (*giustificato motivo*).

Dismissal is not permitted, except for 'just cause', or where the company ceases to trade, in the following circumstances:

- For one year following a female employee's marriage.
- For one year after a woman has given birth or during pregnancy.
- Following an accident or occupationally caused illness.
- On grounds of sickness (the duration of this protection is governed by collective agreement).
- Military service.
- For a year following service on employee representative bodies.
- Following election to public-service bodies.
- Where employees take part in industrial action.

Fair and unfair dismissal

Individual dismissals are regulated by two main pieces of legislation, law 604/1966 and the Workers' Statute 1970. Law 604 was amended in May 1990: the change resolved some of the previous conflicts between different items of legislation on termination, and extended dismissal protection to all employees. It states that employees cannot be dismissed except for 'just cause' as defined by article 2119 of the Civil Code or for 'justified motive'. The burden of proof lies with the employer.

'Just cause' is defined as a grave omission committed by either party which, when evaluated both subjectively and objectively, constitutes a serious and irremediable failure to fulfil the contract. Whether such a breach has occurred would normally have to be determined ultimately by a court, taking all relevant factors into account. Generally speaking, however, omissions on the part of the employee often relate to their obligation of loyalty, diligence and obedience; those on the part of the employer could relate to the obligation to pay the employee, or grant leave. 'Just cause' entitles either party to withdraw from the contract without notice. An employer committing such a breach of contract will also be required to pay the necessary penalties, set out below.

'Justified motive' is defined as the obvious failure of the employee to fulfil contractual obligations *or* reasons inherent in the production process, the organization of work or the smooth running of the undertaking. That is, it covers dismissal related to the conduct or person of the employee (termed 'subjective' grounds) and redundancies (termed 'objective grounds'). An employer may dismiss an employee provided 'justified motive' can be proved and the correct procedures are followed.

Individual termination procedure

There are no specific procedures for termination motivated by 'just cause'. Dismissal for 'justified motive', however, must be in writing. The employee is entitled to ask the reason for dismissal within fifteen days, and the employer must reply within seven. In cases of disciplinary dismissals the conciliation procedures laid down in article 7 of the Workers' Statute must be followed. The article states that an employee can request, via their trade union, that their case should be heard by a conciliation and arbitration tribunal, provided the request is made within twenty days of the employee receiving the written grounds of the dismissal. The tribunal operates under the auspices of the provincial labour authority, and is composed of a representative of each party and a third member agreed by both sides: if there is no agreement the third member is nominated by the labour authorities.

Period of notice

According to article 2118 of the Civil Code either party may terminate a contract of unspecified duration by giving the required notice in the manner specified by existing regulations or by custom and practice or according to equity. Should either party fail to give the required notice they become liable for a payment equal to the remuneration which would have been paid during the period of notice. The employer must also make such a payment in lieu of notice in cases where the employee dies in service. The length of periods of notice is governed largely by collective agreement at national industry level and varies according to sector, category of employee and, often, length of service.

For example, an executive in industry with less than two years' service is entitled to a notice period of eight months. This rises by an additional half a month with every extra year of service up to an overall maximum of twelve months. In the commercial sector, on the other hand, a qualified employee with more than ten years' service is entitled to a notice period of twenty calendar days. Under the agreement, periods of notice are the same for employer and employee, and failure to observe notice by either side renders the party liable to a sum equivalent to salary during the period of notice.

Recourse to law, and settlements

Where a dismissal is considered unfair it must be challenged in writing within sixty days of receiving notice of dismissal. Penalties for unfair dismissal vary, depending on company size and structure. Employers employing more than fifteen employees (or more than five if in agriculture) in each establishment, branch, office or autonomous department, employers employing more than fifteen employees (or more than five if in agriculture) in the same commune, even if they are working in smaller production units, and employers employing more than sixty workers wherever located are liable to:

- Reinstatement of the employee and payment of damages equal to a minimum of five months' pay.
- The employee can refuse reinstatement and request payment of damages equal to fifteen months' pay.
- If the employer invites the employee to return to work, and the employee does not take up the offer within thirty days, the contract is automatically terminated.

In the case of employers employing up to fifteen employees in each production unit (and up to five employees if in agriculture) and employers employing up to sixty employees wherever they are located:

- The case cannot come before the *pretore* (magistrate) unless conciliation has been requested beforehand according to the procedures laid down in collective agreements or in articles 410 and 411 of the Civil Code.
- If attempts at conciliation fail, either party can request within twenty days that the matter be referred to a college of arbitration (*collegio arbitrale*).
- If the matter is still unresolved and the court finds against the dismissal the employer must reinstate the employee within three days or pay damages of between a minimum of two and a half months' pay and a maximum of fourteen months' pay.

  The following maximum damages apply: up to six months' pay, depending on numbers employed, length of service, and the behaviour and circumstances of each party, but up to ten months' pay if the employee has more than ten years' service and where the employer employs more than fifteen workers. If the employee has more than twenty years' service, damages can extend up to fourteen months' pay.

Severance payments

There are no statutory payments specifically in the event of redundancy. In many instances, redundant employees receive a continuing income from the CIG fund (see above), which pays 80 per cent of previous net earnings for up to two years. However, all employees are entitled to a severance payment (*trattamento di fine rapporto*), paid irrespective of the grounds for termination and based on length of service. For white-collar employees the payment is calculated on the basis of one month's pay for each year of service. Blue-collar employees receive less, ranging from fifteen thirtieths to twenty-two thirtieths of one month's pay for each year of service.

Final formalities

On termination of an employee's contract of employment the employer must:

- Pay all sums due to the employee, including all indemnities, remuneration and other benefits or entitlements. Where there is a dispute over the amount to be paid the employer must pay any sums that are not in dispute. Compromise settlements are not allowable.
- Return all documents to the employee promptly at the proper time and make all the necessary entries in the employee's employment documents (*libretto di lavoro*).
- Where employment documents (*libretto di lavoro*) are not obligatory the employer must provide a certificate indicating the duration of the employee's employment with the firm and a description of duties.
- Provide the employee with a copy of the income declaration form which has been sent to the National Social Insurance Institute (INPS).
- Notify the local employment office within five days of the termination of employment.
- Notify INPS of the termination before the fifth day of the month following or, in the case of foreigners, within twenty-four hours, indicating their destination.
- Notify the relevant sickness benefit organization via the monthly return.

Appendix

Ministry of Labour and Social Security
(Ministero del Lavoro e della Previdenza Sociale)
Via Flavia 6
00187 Roma
tel. +39 6 4683

Employment Directorate
(Direzione Generale dei Rapporti di Lavoro)
as above

AIDP (Italian personnel managers' association)
Via Cornalia 19
20124 Milano
tel. +39 2 67 09 558

Confindustria (central Italian employers' organization)
Viale dell'Astronomia 30
0100 Roma
tel. +39 6 59031

BSI Italia Sr1
Via Monteverdi 5
20131 Milano
tel. +39 2 2951 0217
fax +39 2 2951 0168

Acknowledgement

The authors are indebted to BSI Italia Sr1.

8

The Netherlands

Terms and conditions of employment in the Netherlands are widely determined by collective agreements, which cover around 80 per cent of employees. Although industry agreements remain dominant, there was a shift towards company-level agreements during the 1980s, and such agreements covered an estimated 550,000 workers out of a work force of some 5·8 million. Collective provisions take precedence over statute law where they are more favourable to the employee. Legal requirements on termination of contract are unusual in that ordinary dismissals with notice, including redundancies, must be submitted, backed up by evidence, for authorization by the local labour authorities before notice can be given.

Contracts of employment

The contract of employment (*arbeidsovereenkomst*), described in article 1637 of the Civil Code, is the primary form of contract for services recognized in Dutch law. Individual contracts of employment are subject to statute law and the terms of any collective agreement: should a provision in an individual contract provide for worse provision than has been collectively agreed, the corresponding agreed term will apply.

No contract of employment can be concluded with a person aged under 18 years without the verbal or written agreement of their guardian, given in the presence of the employer or the employer's authorized representative.

Form of contract

A contract of employment may be written or oral, although some provisions between the parties have to be set out in writing. They include:

- Contractual periods of notice.
- Restrictive covenants.
- Disciplinary regulations.

Probationary period

A probationary period may be agreed, subject to certain conditions. Although there is no requirement for a probationary period to be specified in writing, the stipulation does occur in some collective agreements. The maximum permitted duration for a probationary period is two months, and the same term must apply for both employee and employer. There is no prescribed minimum period. A probationary period may not be renewed, unless the employer proposes to offer the employee training in a new occupation.

Restrictive covenants

According to article 1637 of the Civil Code, any provision which restrains an employee's activities after termination of the contract must be set out in writing, although provisions in a company's own rule book for personnel would suffice. The effect of the provisions must not exceed what is required to match the legitimate interest of the employer in protecting areas of the business. Both the length of time for a restrictive covenant and the spatial extent must be specified. Breaches of the provision render the employee liable to financial penalties payable to the employer.

Confidentiality

Confidentiality is an implicit duty of the employee, although this may be defined more specifically by the employer. Breaches of confidentiality could warrant dismissal without notice.

Fixed-term contracts

A fixed-term contract may be agreed either for an agreed duration or in order to allow a specified task to be accomplished. This latter may include replacing a named absent employee. Current legislation states that a fixed-term contract which has been extended is presumed to have been renewed for a term equal in length to the original term, under the same terms and conditions. There is no statutory minimum or maximum period.

Working time

Hours of work

The Labour Act (*Arbeidswet*) of 1919 restricts the length of the working day and/or the working week as follows:

- In factories and offices: eight and a half hours per day/forty-eight hours per week.
- In shops: nine hours per day/forty-eight hours per week.

Exemptions may be granted by the labour inspectorate under special circumstances.

Normal working days are Monday to Saturday, inclusive; Sunday working (that is, work between midnight on Saturday and midnight on Sunday) is generally prohibited, although the law does provide for some exemptions to cover work in public services such as transport, catering, health, emergency work of various kinds, maintenance and repair work. Exemptions for those aged under 18 are allowed only in exceptional cases. Employees are entitled to thirty-six consecutive hours off each week, and at least one Sunday off in every two.

Night work

Night work is also generally prohibited, but may be introduced with authorization from the labour inspectorate. Women's night work in industry has been possible since the enactment of legislation in mid-1985.

Agreed working hours

Statutory regulations have been superseded by the provisions of collective agreements in virtually all industries. The average working week laid down by collective agreements is now forty hours or less, although the actual working week may be longer, especially for shift workers. Where this is agreed, any extra weekly or daily hours worked are compensated for by additional days off or free shifts (*roostervrije dagen*).

Major agreements, such as those for the construction, engineering, textile and banking sectors, all specify a standard eight-hour day over a five-day week, giving a forty-hour working week for day workers.

In banking, agreed annual working time in 1990 was 1,824 hours, equivalent to a thirty-five-hour week. Employees work forty hours and take additional time off. In engineering, standard annual working time is 1,848 hours; actual 'basic annual working time' consists of these standard hours less twenty-six half-days off, to be taken throughout the year, equal to 1,744 hours annually or eight hours a day for days actually worked.

Rest periods

Young persons are not allowed to work for more than eight hours a day. They must have an unbroken period of at least twelve hours off duty after a normal working day, and one unbroken period of at least sixty hours (or two unbroken periods, each of at least thirty-six hours), including a Saturday or a Sunday, during any consecutive seven-day period. A young person must also be allowed to have a break of at least half an hour after each four and a half hours' work.

Overtime

The 1919 Labour Act also regulates overtime working. For employees over 18, the law specifies that both male and female employees may, with the authorization of the labour inspectorate, work up to sixty-two hours in any one week, based on the working week under the Act of forty-eight hours. Authorizations are in practice given only for up to fifty-two hours, or twelve hours overtime, in any one week.

Employees under the age of 18 are not allowed to work overtime, and collective agreements typically provide that such overtime is defined in terms of work beyond agreed hours rather than statutory overtime.

Time off

Annual leave

Statutory provisions, laid down in articles 1638bb–1638mm of the Civil Code, state that an employer must grant employees paid annual leave which is equal to at least four times the number of days they regularly work each week. (Employees working a five-day week will therefore be entitled to twenty days' annual leave). The same provisions apply, pro rata, to part-time employees. Employees with less than one year's service are entitled to leave pro rata with length of service.

An employer often specifies a holiday period, usually between April and September, during which at least ten days must be taken together. Similarly, collective agreements often stipulate the number of days' leave which may or must be taken together. The remaining days are then referred to as 'floating' days (*snipperdagen*). A company or sector may, additionally, fix a number of these *snipperdagen*, for instance to cover the shut-down between Christmas and New Year, or to bridge a gap between a public holiday and a weekend.

Collective agreements generally improve on the legal minima. The terms of collective agreements usually specify a basic holiday entitlement for all employees, with actual length of leave often depending on age. In banking, for example, leave entitlement depends on age: twenty-five days are granted for employees aged up to 34 years, with twenty-seven days between 35 and 44, twenty-nine days between 45 and 54, and thirty days between 55 and 64 years. In engineering the basic entitlement is twenty-three days, with extra leave for employees aged over 50, and a maximum of thirty-two days at the age of 64.

Many collective agreements also specify additional entitlement for employees in certain categories such as young workers and those approaching retirement, as well as providing extra days according to length of service.

Public holidays

Provision for paid public holidays is regulated in collective agreements, which generally grant time off for religious holidays and local festivals. There are customarily eight recognized religious holidays.

Sick leave

Employees are entitled to state sickness benefit from the third day of illness for a maximum period of fifty-two weeks, after which a disability benefit is payable. Sickness benefit is 70 per cent of earnings up to a predetermined ceiling of, currently, just over £20,000. The employer is responsible for payment of wages during the first three days' sickness. Some employers make up any lost income to employees by bridging the difference between benefit and net pay, and cover additional employee benefits during long-term disability with supplementary insurance. For example, agreements at Unilever-Nederland, and in the banking and engineering industries, provide for full pay for the first year of sickness, provided the employee has not caused his or her incapacity through negligence.

Termination of contract on grounds of sickness is not permitted, unless the illness has lasted for two years, or if the employee is still working a probationary period.

Maternity leave

During 1990 maternity leave was extended from twelve to sixteen weeks, bringing Dutch legislation into line with practice elsewhere in Europe. Four weeks of the leave must be taken before the expected date of confinement. State benefit at 100 per cent of former earnings, up to the daily earnings ceiling for contributions, is payable. Collective agreements often add a further two to four weeks to the statutory minimum. An employee may also extend the statutory eight weeks after confinement by a further eight weeks, on half pay. The additional leave must be taken during the first year following the birth of a child — not necessarily immediately after birth — subject to agreement between the employee and the employer.

Some company agreements provide for extended unpaid maternity leave of up to two years.

Parental leave

From 1 January 1991 men and women with at least one year's service who have young children — either born to them or adopted — under 4 years of age are eligible for parental leave. It is available on an unpaid part-time basis, with a minimum working week of twenty hours, for a maximum period of six months. Leave can be taken at any time until the child starts school (6 years of age). A statutory right to parental leave exists with regard to each child, but in the case of twins leave is granted only once.

Educational leave

Most collective agreements contain provisions relating to educational or training leave, often on full pay. Such leave may be a few days' training or a longer, part-time course for a recognized qualification. Such leave may, depending on the collective agreement, include leave to attend trade union courses or pre-retirement courses.

Other rights

Legislation on leave of absence is laid down in article 1638c of the Civil Code. It provides for paid time off to be granted to employees over and above their normal annual leave entitlement for personal reasons and family events.

The law provides for one day's paid leave for family obligations, two days' paid leave for birth or marriage, and up to four days' paid leave in the event of a death. The same article also provides for a short and 'reasonable' (unspecified) period of paid leave to be granted to employees who need to fulfil legal obligations.

Collective agreements specify, in addition to these statutory entitlements, more detailed provisions for paid leave of absence for a variety of personal reasons. Such leave is usually referred to as 'short leave' (*kort verzuim*) or 'authorized absence' (*geoorloofd verzuim*). Examples include, from the engineering industry: one day off for the birth of a child (for male employees), for employees' twenty-fifth or fortieth wedding anniversaries, for the death of a relative apart from spouse, parent or child; two days off on the employee's marriage, or the death of a parent; four days off for the death of a spouse or a dependent child.

In addition, employees may be given, by agreement, 'reasonable' time off for dental or medical appointments, for job-hunting (if an employee has been made redundant) and other personal needs.

Retirement

State pensions

Retirement under the state scheme is at 65 for men and women. There is no provision under the state scheme for retiring earlier or later than the normal retirement age. Compulsory contributions are paid by the employee alone until the age of 65. As from 1 January 1991 they are set at 14·05 per cent of gross earnings subject to an annual earnings ceiling of Dfl 42,966 (£13,060) p.a. In general a net pension for a married person is equal to 50 per cent of the net minimum wage if both partners are older than 65; 70 per cent of the net minimum wage if one partner is under 65; 70 per cent of the net minimum wage is payable for single people without children, and 90 per cent for a single parent.

The state pension provision per month from 1 January 1991 is as follows:

- Married persons, with a partner above 65, Dfl 938.11 gross (£285).
- Single persons, Dfl 1,346.90 gross (£409).
- Single persons with a child under 18, Dfl 1,687.32 gross (£512).

Statutory holiday pay, at the rate of 7·5 per cent of annual salary, is added to these amounts.

State pensions may be supplemented by either mandatory occupational or sectoral schemes and/or company pension plans.

Sectoral schemes

Mandatory sectoral schemes exist in a number of industries and for a variety of occupations. The schemes differ substantially, as regards both contribution arrangements and benefits provided, which range from a simple pension to pensions plus widows' and children's benefits. Schemes must be funded. A company may be exempted from a mandatory scheme if it offers company benefits at least as favourable as the relevant sectoral scheme. A company may also request exemption if its own scheme has been in operation for at least six months before being required to participate in a mandatory scheme.

Company pension schemes

As with industry schemes, non-mandatory company pension and other post-retirement or death benefit plans must be funded, either through a company scheme meeting a set of prescribed criteria or through an external pension fund.

Funds must be independent, have Ministry of Social Affairs approval, and in the case of company-funded or insurance-financed pensions are subject to employee consultation and representation. No more than 10 per cent of the fund can be invested in the employer's business.

Early retirement

Early retirement provisions, which currently cover an estimated 1·5 million employees, do not exist under the state scheme, but are regulated by collective agreement and known as VUTs (*vervroegd uittreden*). In most cases, retirement can take place between 60 and 65 with a typical service qualification of ten years in the industry or firm, depending on the type of collective agreement. Benefits are often jointly financed by employer and employee contributions, and disbursed on a pay-as-you-go basis, providing for a bridging pension of 70–80 per cent of earnings until normal retirement age is reached. A minimum pension provision is often specified. In 1991 the average combined contribution rate was 1·4 per cent, with employers paying on average 1·03 per cent and employees 0·37 per cent.

Equal treatment

Sex discrimination

Legislation on equality between the sexes is based on three EC directives: 75/117/EEC, on equal pay for men and women; 76/207/EEC, on the equal treatment of men and women at work; and 79/7/EEC, on the equal treatment of men and women in matters of social security. The first two directives were given the force of law in the Netherlands by three pieces of legislation: the Equal Pay Act (*wet gelijk loon voor vrouwen en mannen*) which came into force on 2 April 1975, the Equal Treatment Act (*wet gelijke behandeling van mannen en vrouwen*) which came into force on 15 March 1980, and the Equal Treatment Act for public-sector employees (*wet gelijke behandeling van mannen en vrouwen in de burgerlijke openbare dienst*), which came into force on 19 July 1980. The legislation on equal treatment was revised in 1989; the three Acts were integrated into one, the Equal Treatment Act, which came into force on 1 July 1989. The part of the Act which refers to employment contracts is included in the Civil Code under article 1637ij.

The Equal Treatment Act. The Act stipulates that no distinction must be made either directly or indirectly, between men and women. The Act specifies that all elements of pay — basic rates, premiums, supplements and bonuses — must be calculated on the basis of equal criteria for both men and women, provided they are doing work of equal value. The Act also sets rules for career guidance, training, recruitment (including job advertisements), conditions of work, promotion and entering into or terminating a contract of employment. In the case of employment contracts, it is also illegal to make any distinction between men and women on grounds of marital status or family situation.

The Act allows a few exemptions:

- Certain types of employment, such as entertainment or in the Church, are treated as 'special cases'.
- To preserve decency and privacy. This can arise if a job involves physical contact between the sexes which might cause embarrassment or offence (e.g. staff in old people's homes).
- Special safeguards in connection with pregnancy and childbirth.
- Positive discrimination in favour of women employees, in order to promote equality of opportunity in occupations where women are underrepresented.

All employees are covered by the Act; it applies not only to employment situations with a formal contract of employment, but also to training and voluntary work. The professions are now also covered.

Equal Treatment Commission. The Equal Treatment Commission was established in 1975. Because there were two Acts on equal treatment prior to 1989, there were also two separate commissions, one for the private and one for the public sector. Where the law was revised in 1989 a new Equal Treatment Commission was set up, replacing the two previous ones. The commission has eleven members, each selected for a period of four years.

The commission has both an investigative and an advisory role. Its main task is to examine alleged cases of discrimination and set up a formal investigation of the Equal Treatment Act. It produces a series of publications which give guidelines on, among other things, job advertisements.

Grievance procedure. An employee who feels that he or she has been discriminated against on grounds of sex can take the matter up — in writing — directly with the commission, which examines the case (the preliminary investigation), and if, in the opinion of the commission, there is sufficient evidence of genuine sexual discrimination the commission can commence a formal investigation and give its verdict. The commission cannot, however, impose any legal sanctions or compensation payments, which must be determined by a court.

The 1989 revised Act not only applies to individual companies but also gives trade unions and works councils the right to present a complaint on behalf of a group of employees (*groepsactie*) and, where necessary, to start legal proceedings.

The grievance procedure through the Equal Treatment Commission is free of charge.

Disabled workers

According to the legislation on disabled workers (*wet arbeid gehandicapte werknemers*, or WAGW) employers are obliged to employ a quota of disabled workers of between 3 and 7 per cent of their total work force, depending on the sector and the company's particular type of business. In reality, however, quotas of this size have not been achieved anywhere. A sample collected by the Ministry of Labour in 1990 showed that an average proportion of disabled employees in a firm was just 2·19 per cent. Employers do not have the right to release themselves from their obligation through a payment.

Ethnic minorities

Racial discrimination is formally forbidden by law (the Penal Code). In addition, separate legislation governs non-national employees, guaranteeing terms and conditions equal to those enjoyed by Dutch nationals.

Work environment

Framework legislation on the working environment, the 'Arbo-Wet' (Working Conditions Act) was originally passed in 1980 and introduced in phases, with the last phase entering into force in October 1990. This law lays down broad principles which can be augmented by ministerial order.

Employers have the following general duties:

- To promote and ensure the welfare, health and safety of employees.
- To give adequate information to employees on working conditions and safety.
- To assign tasks appropriate to an employee's age, and physical and mental capacities.
- To avoid monotonous work, where possible.
- To organize day-to-day operations to ensure the best possible working conditions and make changes in the light of scientific or technological developments.
- In companies with 100 employees or more, to report annually on working conditions.

Employees may sue their employer for an alleged breach of these legal obligations and may ask the labour inspectorate, the works council, where such exists, the health and safety committee or the trade union to intervene. Should an employee consider that their present work poses a serious and immediate physical danger, they can stop work without loss of pay, provided the employer is notified immediately, and that there was no time to notify the labour inspectorate directly.

Works councils are normally responsible for overseeing and promoting safety at work, but this task may also be assumed by a specifically established health and safety committee where no works council exists. Health and safety committees have a broad body of rights, including the right to consult outside experts on safety matters at the employer's expense. Employers, safety officers and company doctors must inform committees on safety issues.

Industrial establishments with more than 500 employees must set up a safety department, with specialist staff. This requirement may also apply to other establishments, depending on the nature of their operations. The main responsibilities of the department are to promote occupational health and safety, prevent accidents, and participate in the formulation of safety plans and mandatory reports. Establishments with more than 750 employees are required to set up medical services, or guarantee access through a joint service. The responsiblities of the medical department include:

- Examination of newly engaged employees or when required under collective agreements.
- Monitoring of and, if appropriate, making recommendations on occupational disease.
- The organization of first-aid services.
- The provision of treatment, subject to the agreement of an employee's own doctor and the consent of the employee.

Individual termination of contract

An indefinite contract of employment may be terminated by the employer either summarily or by giving the required notice, depending on the grounds of the termination. In the case of individual dismissals there is no difference in law between termination for reasons related to the person or conduct of the employee and termination on economic grounds. (Collective dismissals, and the associated notification and consultation requirements, are dealt with in volume 3 of the European Management Guides series). No notification of work-force representatives is required for an individual dismissal.

Termination procedure

Before giving notice to terminate a contract of employment the employer must first inform the employee and apply for official permission to end the employment relation. Except for specified exceptions, the termination of a contract of employment requires authorization from the regional office of the department of employment, the GAB (*gewestelijk arbeidsbureau*). No authorization is needed:

- During or at the end of a two-month probationary period.
- Where termination is by mutual agreement.
- When a fixed-term contract expires.
- For summary dismissal on grounds of serious misconduct (see below).
- For directors of a public company (BV) or limited liability company (NV).
- If the contract is terminated by a district court on serious grounds.

Employers must demonstrate the grounds for the termination, and the GAB may also set conditions on the ending of the contract and its aftermath.

Application must be in writing, with details of the employee and the reasons for the dismissal. The GAB forwards a copy of the application to the employee within three days. The employee has fourteen days to respond to the proposed termination. The application, together with any additional information from the parties, is considered by the joint consultative machinery of the regional employment office, and a final decision is made by the director. Once authorization has been issued, the employer may proceed to give due notice for a period of up to two months.

Authorization may be granted:

- For operational and economic reasons.
- In the event of poor performance, and the unlikelihood of a successful transfer within the enterprise or an improvement.
- Where the employee has been absent on grounds of illness for at least two years.

If the termination is on economic grounds the employer must justify the dismissals by making evidence available indicating the financial state of the enterprise, the effects of the redundancy, and the reasoning behind the choice of employee. (As with dismissal on individual grounds, this evidence is available to the employee.) Up to nineteen employees over a three-month period may be dismissed on economic grounds without triggering the consultation provisions required in the event of collective dismissal (dealt with in volume 3 of the European Management Guides series).

Period of notice

A number of factors have to be taken into account when fixing the minimum period of notice in the case of indefinite contracts:

- *Frequency of pay*. Statutory: the minimum period of notice must equal at least the interval between two successive pay days, normally one week or one month. Collective agreements and written contracts may diverge from this, provided the change benefits the employee and the period of notice is no longer than six months.
- *Length of service* determines additional notice. Notice must not be less than one week for each year of service, but must be no longer than thirteen weeks. (Service is calculated from the age of majority — eighteen or twenty-one, depending on marital status.)
- *Age*. For employees over 45 years of age, if the period of notice is calculated using the 'length of service' criteria, it must be increased by one week for every complete year of service after the age of 45, subject to a maximum of an additional thirteen weeks, and an overall maximum of twenty-six weeks. The employer must give at least thirteen weeks' notice after one year's service in the case of dismissing employees aged between 50 and 65.

Although it is not compulsory, notice should be given in writing. Moreover, collective agreements often stipulate that notice must be in writing.

Summary dismissal

Dismissal without notice is permitted for an urgent reason, usually related to misconduct on the part of the employee, where it would be deemed unreasonable for the employer to have to allow the employee to work out their notice. Instances would include:

- Malicious deception on the part of the employee on hiring.
- Violence, theft, drunkenness.
- Serious negligence or incompetence.
- Refusal to accept the employer's instructions.

Protected employees

The following groups of employees may not be dismissed, except for summary dismissal:

- Employees absent through sickness (for the first two years), unless the employee is working a probationary period.
- Pregnant women.
- Women on maternity leave.
- Works council or health and safety committee members. Should an employer seek to dismiss a work-force representative through termination with notice for up to two years after their period of office, the employer's application must pass through a more rigorous test of whether authorization is warranted than for an ordinary employee, including submission of the application to a judge.
- Employees doing military service.

Dismissal in any of these cases is not legally void, but could render the employer liable to pay damages and reinstate the dismissed employee.

Severance payments

There is no statutory provision for either a severance indemnity or for redundancy payments. Such payments may, however, be required under collective agreements.

Appendix

The Ministry of Social Affairs and Employment
(Ministerie van Sociale Zaken en Werkgelegenheid)
Postbus 20801
2600 EV's-Gravenhage
tel. +31 70 371 5911
fax +31 70 371 4555

Dutch Association of Personnel Managers
(Nederlandse Vereniging van Personeelsbeleid NVP)
Catharijnensingel 53
Postbus 19124
3501 DC Utrecht
tel. +31 30 367101 (Membership enquiries), 367155 (training), 367137 (administration)

Social and Economic Council
(Commissie Ontwikkeling
Bedrijven van de Sociaal-
Economische Raad, COB/SER)
Postbus 90405
2509 LK 's-Gravenhage
tel. +31 70 3499 499

Labour Foundation
(Stichting van de Arbeid)
Postbus 90405
2509 LK 's-Gravenhage
tel. +31 70 349 9648

Netherlands–British Chamber of
Commerce
Holland Trade House
Bezuidenhoutseweg 181
2594 AH 's-Gravenhage
tel. +31 70 478881
fax. +31 70 477975

9

Portugal

The development of employment law during the late 1980s has seen a number of moves away from the provisions established in the wake of the 1974 revolution. The most prominent recent step has been the contentious package of five decrees on contracts of employment and dismissals passed in 1989. Further proposals aimed at amending current laws on working time, leave, probationary periods, transfer and dismissal are under consideration.

The present constitution was promulgated on 2 April 1976, with additions anchoring certain social rights in 1982 and 1989. The constitution guarantees the right to work (article 59), and obliges the government to take steps to pursue full employment, equality of opportunity and the cultural, technical and professional education and training of employees.

Contracts of employment

The bases of Portuguese law on the individual contract of employment are set out in decree law 49408 (24 November 1969).

As in the United Kingdom, Portuguese law recognizes a difference between a contract of employment (*contrato de trabalho*), under which an individual enters into a commitment to work under the supervision of an employer, and a contract for services (*contrato de prestação*), in which one party agrees to furnish the other with specific and completed goods or services.

A contract of employment entails a number of rights and obligations for both employer and employee. For example, the employer is required to provide appropriate pay and working conditions, and the employee is obliged to exercise loyalty, zeal and diligence towards the employer.

Specifically, an employer may not seek to prevent employees from exercising their individual rights, exert pressure on employees which might adversely affect their working conditions, or lower employees' pay or transfer them to a lower career level except under circumstances — usually temporary — provided for in the law. For example, if the individual contract specifies a particular level of pay and benefits, these must be maintained. Any change in the terms of the individual contract must be by mutual agreement and must be submitted to the Ministry of Labour for authorization.

Contract of employment for an indefinite term

A contract of employment for an indefinite term does not require a particular form: it may be oral, in writing, explicit or implied. However, any unusual terms, such as a restrictive covenant, have to be in writing. Moreover a contract must be in writing should either party insist.

The conclusion of a contract of employment presupposes that an offer of employment has been made and accepted (on recruitment procedures and practices, see volume 1 in the European Management Guides series). The parties have to be in agreement as to the essentials of the contract, and specifically that the employee agrees to work for the employer and that the employer agrees to pay the employee. Should disagreements arise as to the interpretation of a contract, then the burden of proof rests with the party seeking to obtain their rights under it. The presentation by the employee of working papers (such as income tax documents), and the deduction of tax and social security contributions by the employer, are deemed to denote that a contract of employment has been concluded. Employees are presumed to have accepted the terms of employment provided by an employer if they do not object within thirty days. Proof of a contract may also be obtained through a witness.

A contract may become null and void under various circumstances, including:

- A defect in the consent by one of the parties to the agreement, such as conclusion by a minor without the approval of a parent or guardian.
- Fraud.
- Constraint.
- Inability to understand the full meaning of the acceptance of the contract.
- If the agreement is intended to fulfil an illegal purpose.

Employers and employees are free to settle the terms of a contract of employment, subject to statute law and the provisions of any applicable collective agreement. Individual terms may diverge from either if they are more favourable to the employee, but an employer may not arbitrarily discriminate between employees as regards their terms and conditions, nor treat employees more or less favourably on grounds of sex, religion, political affiliation or union membership.

Probationary period

A probationary period may be agreed, which typically lasts sixty days, to be counted from the beginning of the contract's application. However, this sixty-day period may be reduced or even eliminated by the agreement of a specific provision in the contract or through an instrument of collective regulation. Conversely the probationary period may be extended up to a maximum of six months, either through the individual contract or through an instrument of collective regulation, if the employment entails technically complex tasks for which a sixty-day period would not be sufficient to judge the competence of the employee.

Under a contract for an indefinite term, employees can be dismissed only on grounds to be established under a prescribed disciplinary procedure (see below), in the event of collective redundancy, or where the job ceases to exist on economic, technological or structural grounds.

Special probationary periods apply in the case of fixed-term contracts: these are set out in the section on fixed-term contracts below.

Restrictive covenants

Any limitations on an employee's ability to work freely for another employer after the contract has ended must also be clearly related to potential harm to the previous employer's business, and the employee must be remunerated during the period of restriction. Such a restraint may be concluded only for a period of up to three years.

An employer may oblige an employee to repay training costs, if the employee had entered into a commitment to work for a specified period, or subtract a sum for training from any compensation for a restrictive covenant.

Period of notice

There are no notice periods as such for the termination of an individual contract, as dismissal on grounds related to the conduct of the employee must pass through a prescribed disciplinary and/or judicial appeal procedure, during which employment continues. (See termination of contract, below, which also sets out the employee's obligations.)

Fixed-term contracts

The current law on fixed-term contracts was introduced in February 1989 (legislative decree 64-A/89). Fixed-term contracts may be concluded only for:

- The temporary replacement of a worker who, for any reason, is prevented from rendering service or against whom legal proceedings for lawful dismissal are pending.
- A temporary or exceptional increase in the enterprise's activities.
- Seasonal work, and occasional tasks or specified services of a defined and non-enduring character.
- The launching of a new activity of uncertain duration or the commencement of activity in an enterprise.
- The execution, direction and supervision of activity in civil construction works, public works, industrial building and repairs.
- The development of projects, including planning, research, direction and supervision, which are not a customary part of the employer's activity.
- Hiring workers for their first job or the long-term unemployed.

Fixed-term contracts must be in writing, must be signed by both parties and must set out the employee's skill or qualification, remuneration, the agreed duration and the reason for the contract. Failure to do so could mean that the contract might be deemed to have been concluded for an indefinite term.

In the case of temporary replacement, seasonal work, civil and public works and project development the contract may be concluded for an unspecified term, with the contract ending when the task has been completed.

The probationary period within which either side may withdraw is fifteen days for a contract not exceeding six months, and thirty days for any other fixed-term contract.

A fixed-term contract subject to renewal may not be repeated more than twice, and the total duration of the contracts may not exceed three successive years. In the case of a contract for launching a new activity the maximum length is two years. Employment which exceeds the agreed period and where the renewal has occurred more than twice or gone beyond three successive years will be deemed to be indefinite.

Any employer who does not want to renew a fixed-term contract must notify the employee in writing at least eight days before its expiry. If the contract has not been renewed twice, nor three years have elapsed, then failure to give the proper notification will entail a renewal of the contract for a term at least equal to the period of the first contract. At least three months must have elapsed between the expiry of a fixed-term contract exceeding twelve months and the hiring of a new employee in the same position.

Part-time work

There are no specific regulations governing the use of part-timers. Part-timers enjoy the same basic rights as other employees, and must be paid the proportion of the appropriate full-time remuneration which corresponds to their hours of work. Employers wanting to introduce part-time employment would need to inform the workers' committee in their establishment. Employees with children under 12 may apply to work part-time for a period of up to three years, followed by reinstatement as a full-time employee, under parental protection legislation passed in 1984 and 1985 (laws 4/84 and 136/85).

Working time

Hours of work

Under decree law 409/71 and law 2/91 normal working hours may not exceed eight hours per day and forty-four hours per week in all sectors. For office employees there is a maximum week of forty-two hours. Typical agreed hours in industry are about forty-four, with a prospect of reductions in the future, with thirty-seven and a half to forty hours in banks, insurance companies and airlines.

A break must be provided lasting not less than one hour but not more than two, so that workers do not work longer than five hours continuously. Commercial and industrial enterprises must close for one day a week, unless given authorization, with another regular rest day in most of the major sectors of the economy (industry, commerce, haulage). Employers must by law display a notice indicating the beginning and end of daily work, weekly rest days, and any other closing period.

On application from employers, the labour inspectorate (IGT) may permit a reduction of or exemption from these breaks if the new arrangement is to the benefit of employees, or if there are special warranting circumstances.

The legislation on working hours is currently being revised.

Overtime

Overtime may be required of employees under decree law 42/83, unless there are established grounds to the contrary. Exempt employees include disabled workers, women with children less than ten months old, and minors. The definition of overtime excludes:

- Work performed beyond normal working hours on a normal working day by employees not subject to the usual working schedules.
- Work performed to compensate for any suspension of operations not exceeding forty-eight hours continuously, or interrupted by a rest day or holiday where there is an agreement to that effect between the employer and employees.

Overtime is permitted only if the establishment needs to meet a sudden demand for additional production, but where this does not warrant hiring extra employees on fixed-term contracts, or in the event of *force majeure* to ward off serious impending damage (in which case no limits apply).

Overtime is subject to precisely defined limits. Except in exceptional circumstances it may not exceed 160 hours per year and two per day per worker.

By law the overtime supplement for the first hour must be at least 50 per cent of usual rates, with 75 per cent for additional hours, and 100 per cent on rest days and holidays. Employers and employees are obliged to pay 25 per cent of the additional income from overtime work into the national unemployment benefit fund.

Establishments with more than ten employees must offer the right to paid time off in lieu corresponding to 25 per cent of overtime worked as soon as a number of hours equal to normal daily working hours has been reached. The time off must be taken within thirty days. If overtime is worked on a prescribed weekly rest day, the employee has the right to a paid day off in lieu within the following three days.

Night work

Night work is defined as any work carried out between 20.00 and 07.00. Under the law, it must be paid for at a supplement of at least 25 per cent of usual day rates, a proportion which may be, and usually is, increased by collective agreement.

A collective agreement may specify a shift pattern under which night work may be regarded as any consecutive period of eleven hours, of which at least seven consecutive hours have to fall within the period 20.00–07.00.

Night work for women in industrial establishments may be authorized in the event of *force majeure* or if necessary to prevent avoidable material losses. No special permission is required if women occupy positions of responsibility, technical functions, do not regularly carry out manual work, or are engaged in hygiene or welfare activities. Night work for minors under 16 years is forbidden in industrial establishments, except where work is not of an industrial character and is deemed indispensable for their vocational training.

Workers on continuous or semi-continuous shifts must have a prior medical examination which is repeated annually.

Time off

The law provides for a number of forms of time off, some of which are paid for by the employer, others not. In general the law protects against loss of pay whenever the absence is for reasons beyond the employee's control, such as sickness or accident, medical treatment, family reasons, civic duties, or for approved study. There is also protection for employees carrying out trade union activities.

Annual and public holidays

There are twelve national public holidays a year, plus an additional local holiday. Employees do not have a right to an additional day off if a public holiday falls on a day not normally worked.

Paid annual leave must be at least twenty-one days but no longer than thirty consecutive days per calendar year. Leave must be taken in the calendar year in which entitlement falls due and cannot be carried over unless there is a specific agreement to this effect, with a time limit of 31 March in the following year.

Maternity leave

Maternity leave totals ninety days, sixty days of which must be taken after the birth. Mothers receive an allowance from the social security system equal to their average earnings in the six months preceding the second month before leave commences. Pregnant women are entitled to additional paid time off for any antenatal medical examinations. Nursing mothers are entitled to two one-hour paid breaks per day until the child is one year old. The father may take the leave should the mother die during the period of maternity leave.

Paid leave of absence

Justified absence under the law, involving no loss of pay, covers the following circumstances:

- Sickness, accident, the need to carry out legal duties, or attend to close relatives. Sick pay under the state scheme is equal to 65 per cent of average earnings in the six months prior to the second month before the period of illness began, and is payable for 1,095 days in all. Under some collective agreements the employer pays the difference between the state benefit and net remuneration. Any period of sickness extending for more than 1,095 days will lead to the employee being retired on grounds of disability unless they return to work.
- Death of a spouse or very close relative: five consecutive days' leave.

- Death of a direct relative or cohabitee: two days' consecutive leave.
- Marriage: eleven consecutive days.

Collective agreements may specify extra leave for other family events.

Family commitments

Assistance to the family is granted in the form of carers' leave of up to thirty days to attend to sick dependent children under 10 years of age. Leave may be extended to cover a full period of hospitalization. The state sickness benefit is payable, unless better provision is offered in a collective agreement. Employees are also entitled to up to fifteen days' leave a year to attend to a sick spouse, parent or a child over 10 years of age, although in this case no state benefit is payable.

Unpaid leave of absence

Employees may apply for unpaid leave, which, if granted, counts towards length of service. (It may be used for educational purposes, or for parental duties.)

Educational leave

Any employees continuing their education are entitled to time off without loss of pay during examinations or other assessment of two days for a written exam, or two days for an oral exam, one of the days being for the actual exam and the other for the preceding day. A dispensation of up to six hours' paid time off a week can be granted for attending classes if the classes clash with working hours.

Retirement

There is no statutory mandatory retirement age, but normal retirement age is 65 for men and 62 for women. Many companies have a common retirement age for men and women. Retirement is voluntary. An employee who is forced to retire will be deemed to have been dismissed without 'just cause', and the employer liable for penalties for unlawful dismissal. Civil servants may be required to retire at 70, and are allowed to retire with thirty-six years' service.

Should an employee still be working thirty days after the date which both employer and employee have recognized as that individual's retirement date, then the contract of employment will become a fixed-term one for six months, renewable, which the employer may terminate with sixty days' notice, or the employee with fifteen days' notice. Once the employee reaches 70 the contract automatically becomes a fixed-term one, which the employer may terminate according to the periods of notice foreseen for fixed-term contracts (see above).

Pensions under the *state* scheme are calculated on the basis of the best five years' earnings within the past ten contribution years, and payments are made fourteen times a year. The level of pension is calculated on the basis of 2·2 per cent of average monthly earnings for each contribution year, with a minimum of 30 per cent of earnings and a ceiling of 80 per cent of the calculated salary. There are additions to pensions where the recipient has a dependent spouse.

Occupational schemes are regulated by legislation passed in 1986 (legislative decree 393/86): the main aim of the law was to provide a framework for the financing of schemes which complement or replace the state system, and in particular to reduce the pay-as-you-go element. As from 1 January 1987 all new occupational schemes must be funded, and tax incentives are available to encourage existing schemes to move on to a funded basis. Funds must be managed either by a 'managing society', constituted under the 1986 legislation, or a life insurance company.

Most private schemes seek to bring an individual's total pension up to around 80 per cent of final earnings, as opposed to the averaging employed under the state scheme, which lowers the value of the benefit where inflation has been high. Few schemes would require employee contributions.

Equal treatment

Equality of opportunity and treatment for women at work, and *mutatis mutandis* for men, is provided for under 1979 legislation (legislative decree 392/79). The law is envisaged as implementing the right to work enshrined in the 1976 constitution, and requires that access to employment should not be obstructed by any discrimination based on sex, whether directly or indirectly, or by reference to marital status or family circumstances. Any law, collective agreement or professional regulation which restricts women's access to employment, or provides for unequal pay for men and women, is deemed to be null and void. Undertakings must ensure that women also enjoy equal access to vocational training once they are in employment. Specifically the law:

- Prohibits discrimination on recruitment through gender-biased job advertisements, or the specification of physical requirements unrelated to the advertised post. Exceptions are permitted in the fashion, arts and entertainment industries where it is essential to the nature of the task.
- Prohibits or limits work which is deemed to imply an actual or potential risk for procreation: these may be reviewed in the light of developing scientific and technical knowledge.
- Requires equal pay for work of equal value done for the same undertaking by men and women. Job evaluation systems must be based on objective criteria that are the same for men and women. Any woman employee who wants to allege discrimination must indicate the comparator male employees: the burden of proof then falls on the firm to show that any differences in pay are based on factors other than sex.
- Requires equality of opportunity in promotion, and protects women against dismissal should they allege that discrimination has taken place. In the absence of proof to the contrary, any dismissal or disciplining of a woman within one year of her alleging discrimination will be held to be unlawful.

Work environment

Draft legislation on industrial hygiene, and health and safety at work has been under discussion since 1990. The law, if passed, would provide a new basic framework for detailed regulations, and would improve machinery for the monitoring of hazards. Some of the main objectives of the law, which must be approved by the tripartite National Council on Health and Safety at Work before progressing through the legislative process, include:

- Identification of substances, agents or processes which require restriction, authorization or monitoring, and the definition of limit values for employee exposures, together with procedures for measuring substances and evaluating the effect of exposures.
- Procedures for reporting accidents at work and occupational disease.
- Greater resources into safety training, and the coordination of the work of existing organizations currently having a responsibility for industrial health and safety.

The two principal organizations engaged in the supervision of health and safety are the General Inspectorate of Labour (IGT), whose sole task is to monitor and control the application of regulations, and the Directorate General of Occupational Safety and Health (see Appendix), which advises on standards and carries out research and education in the prevention of occupational risk.

Discipline and individual termination of contract

Disciplinary procedures and dismissal are closely related where the dismissal is for reasons connected with the employee's conduct. We therefore begin by outlining employers' disciplinary powers.

Employers' disciplinary powers

Employers are granted disciplinary power over employees in their service, under statutory provision (law 49408/69 and 64-A/89) which can be modified by collective agreement. Disciplinary authority can be exercised either directly by the employer or through an employee's superiors as established in a corporate hierarchy. Employers may impose penalties on employees, ranging from reprimand, fines up to a maximum of ten days' pay in any one year (though fines are rarely imposed), suspension without pay of up to twelve days for each penalty and thirty days per year, or reduced pay, to dismissal without compensation.

The law requires that any disciplinary procedure must begin within one year from the alleged offence, otherwise the offence is deemed to have lapsed. Disciplinary proceedings must be carried out within sixty days of any offence becoming known to a superior. Employees may be suspended from work during the course of disciplinary proceedings, but must be paid. No penalty may be decided upon before an accused employee has had an opportunity to give evidence, and failure to hear the accused worker will render the proceedings null and void.

Disciplinary proceedings take the following form:

- Indictment.
- Defence.
- Consultation with employee representatives.
- Decision.

In the indictment stage the employer must clearly identify the accused employee and describe the circumstances of the alleged misconduct. A copy of the 'charge' must be sent to the employee, with a copy for the works committee (*commissão de trabalhadores*) and a copy to the employee's trade union if the employee is a union representative. Should the employer want to dismiss the employee, then he or she must make it clear at this stage of the procedure.

The employee must present a defence to the charge within five working days. Employers must prove their accusations in any way demanded by the employee, unless the employee is merely seeking to delay or thwart proceedings. The burden of proof lies with the employer. Worker representatives are entitled to comment on the proceedings, but their report is not binding.

Employers have a period of thirty days after hearing witnesses and completing established procedures within which to issue a decision.

Termination of contract

A contract of employment can end by:

- Lapse: this includes such factors as the ending of the specified period of a fixed term contract, *force majeure* or the retirement of the employee.
- Revocation by mutual consent.
- Dismissal by the employer with just cause.
- Withdrawal on the part of the employee, with or without material reason.
- Withdrawal by either party during a probationary period.
- The extinction of the position owing to structural, technical or economic factors affecting the enterprise.

Dismissals without just cause are prohibited. Individual dismissal with just cause requires: culpable behaviour by the employee, which is both serious and with grievous consequences (*gravidade e consequencia*), and a close connection between that behaviour and the impossibility of continuing with the employment relation. A dismissal will be deemed to be unlawful if it fails to follow the required disciplinary procedures outlined above; if it is on political, religious or ideological grounds; or if the invoked reason has been found groundless by a court.

Reasons which might warrant a dismissal with just cause include: unwarranted refusal to carry out orders; violent behaviour; an extraordinary decrease in productivity; false statements by the employee about absences (law 64A-89, article 9).

Compensation is payable in the event of unlawful dismissal. This entails:

- The payment of any remuneration due between the period from which the dismissal was effective and the final decision.
- The rehiring of the employee without prejudice, unless the employee opts to receive financial compensation of one month of basic pay for each year (or fraction of a year) of employment, with a minimum of three months' compensation. Compensation may be reduced if the employee has found another job (article 13, law 64-A/89).

Legislation passed in 1989 (decree 64-A/89) introduced the concept of individual dismissal on *economic grounds*, replacing the previous legislative decree dating from 1975 (No. 372). Individual dismissal on economic grounds is permitted if an employee's position ceases to exist because of economic, market, technical or structural factors affecting the enterprise.

In the case of an individual dismissal on economic grounds the employer must demonstrate that:

- The reasons for the termination are not due to factors attributable to the person of the employer or the employee.
- It is impossible for the employment relation to continue because no other appropriate position exists, subject to reasonable retraining.
- The employee is given appropriate compensation.

These stipulations would imply, for example, that no other individual may take over the dismissed employee's job once they have left.

Termination by the employee

The employee may also terminate the contract and claim compensation with 'good reason' should the employer commit a serious breach of the employment contract, if the employee needs to exercise civil duties incompatible with continuing the employment relation, or if the employer changes working conditions 'substantially'. The employee may terminate the employment relation without good reason by giving notice of either thirty days with less than two years' service or sixty days with more than two years' service.

Appendix

Ministry of Labour and Social Security
(Ministério do Emprego e da Segurança Social)
Praça de Londres 2
Lisbon
tel. +351 1 80 44 60

Institute of Employment and Vocational Training
(Instituto do Emprego e Formação Profissional)
Avenue José Malhoa, 11°
Lisbon
tel. +351 1 726 25 36

Association of Portuguese Human Resource Managers
(Associação Portuguesa dos Gestores e Técnicos dos Recursos Humanos), the professional organization of personnel practitioners in Portugal. Founded in 1964, it has 2,000 individual members. Companies may also affiliate.
Avenue do Brasil 194, 7°
1700 Lisbon
tel. +351 1 89 97 66
fax +351 1 80 93 40

Office in Oporto:
Rua Formosa 49, 1°
tel. +351 2 32 32 34
fax +351 2 200 07 64

Confederation of Portuguese Industry
(Confederação da Industria Portuguesa, CIP, the central employer's organization for industry)
Avenue 5 de Outubro 35, 1°
1000 Lisbon
tel. +351 1 54 74 54

Confederation of Portuguese Commerce
(Confederação do Comércio Portuguës), the central employers' organization in trade and commerce
Rua Saraiva de Carvalho, 1°
1000 Lisbon
tel. +351 1 66 85 39

Directorate General of Occupational Safety and Health
(Direcção Geral de Higiene e Segurança do Trabalho)
Avenue Da Republica, 84–5°
1600 Lisbon
tel. +351 1 77 28 22

10

Spain

Current legislation regulating terms and conditions of employment is of comparatively recent origin, as legislators have sought to replace or amend laws dating from the Franco period. The three main sources of contemporary law are: the 1978 constitution, the 1980 Workers' Statute and the 1980 Basic Employment Law. Provision made in collective agreements or agreed individually takes precedence over statute law in setting terms, wherever they provide for more favourable treatment for the employee.

The 1978 constitution guarantees the right of association, and Spain has ratified ILO article 87, providing for employer and employee rights of association. Article 28 of the constitution establishes the right to form or join trade unions and for trade unions to form or join confederations; it outlaws the closed shop and sets financial penalties for infringement. The constitution also establishes the right of professional associations to organize to promote their economic interests.

The 1980 Workers' Statute elaborates on these basic rights. Article 1 details rights of association in the workplace — permitted to all employees except the armed forces and the judiciary. Article 4 reiterates the right of association, and article 61 the right to participate in collective bargaining and collective representation.

The constitution guarantees basic freedom of expression in article 20, which prohibits censorship of any kind, while article 68 of the Workers' Statute gives workers' representatives a specific brief to express their views freely on matters affecting employment conditions, and to circulate freely printed matter 'of employment or social interest'.

Contracts of employment

The employment contract may be verbal or written, and may be freely agreed between an employer and a person aged at least 18 (or a 16–18 year old who lives independently or has parental permission). All contracts for a fixed-term or for carrying out a specific task must be written. Either party in any case may require that a contract must be written. When written, contracts are obliged by statute to indicate:

- The type of employment.
- Whether it is for a given number of hours, or for a specific task.
- The amount and form of pay.
- Working time and rest time.
- Formal arrangements for social security payments.
- An undertaking to abide by labour legislation.
- An indication as to whether sanctions may be applied, and how they are to be applied.
- Information on training provision.
- A statement on whether the contract is permanent, temporary or otherwise.

Unless stated otherwise, contracts are presumed to be permanent.

Executives are not covered by many of the statutory provisions relating to other groups of employees. Their contract may also be verbal or written (necessarily written when an employee is promoted to an executive position). The law relating to these contracts (*real decreto 1,382/85 por el que se regula la relación laboral especial del personal de alta dirección*) determines that the contract should be permanent unless otherwise specified, and that such conditions as working time, holidays and leave are to be negotiated individually.

Probationary period

Probationary periods are determined by statute, and may be defined in the contract of employment. They may be for a maximum of nine months for executives, six months for highly qualified personnel, three months for skilled workers, and two weeks for unskilled workers. During probation, either side may terminate without notice. However, the employee has the same rights as other employees during this period, and the probationary period counts for seniority.

Fixed-term contracts

Fixed-term contracts are a relatively recent development in Spain, and have been generally permitted only since 1984, usually in connection with job creation programmes. In 1989 2,290,448 out of a total of 4,326,500 new contracts were fixed-term. There are nine principal types:

- *Temporary job creation contract* (*contrato temporal de fomento de empleo*). Available to the registered unemployed; with certain restrictions on the employer dismissing other fixed-term employees, it lasts from six months to three years.
- *Work experience contract* (*contrato en prácticas*). Available to the unemployed who have gained qualifications in the past four years; may be from three months to three years, part-time or full-time; they offer the employer social security contribution abatements.
- *Training contract* (*contrato para la formación*). Available to 16–20 year olds for on-the-job training, such contracts may last from three months to three years, with 25–50 per cent of the attendance time allocated to training. Social security contribution reductions and training cost subsidies are available to the employer.
- *Contract for a specific job or service* (*contrato de obras o servicio determinado*). This is available to employ staff to fulfil a certain job or service of limited or uncertain duration. It may last for any period of time, must be written if for more than four weeks, and carries two weeks' notice if it is for more than one year.
- *Casual contract related to production requirements* (*contrato eventual por circunstancias de la producción*). This is available to cope with seasonal or market fluctuations, and may last up to six months in any twelve-month period.
- *Temporary replacement contract* (*contrato de interinidad*). This is available to substitute for employees on temporary leave of absence.
- *Contract for launching a new activity* (*contrato por lanzamiento de nueva actividad*). This contract is designed to allow a new company or an established one to install new production lines, launch a new product or service, or open a new workplace. It may be for six months to three years, and is confined to the registered unemployed.

- *Permanent casual contract* (*contrato de trabajadores fijos y periódicos de carácter discontinuo*). This is available for seasonal or intermittent employment, terminating whenever the work in question is finished — typically in agricultural work. Employees must be summoned in strict order of seniority. Permission must be sought from the employment office not to renew the contract at the start of the next season or promotion.
- *Substitute contract to assist early retirement* (*contrato de sustitución por anticipación de la edad de jubilación*). This contract is designed to allow the registered unemployed to replace a worker aged 64 for a minimum period of one year. There are social security contribution benefits for the employer who converts the contract to a permanent contract on termination.

Part-time work

Part-time contracts became permissible in Spain only in 1981, but have grown steadily in importance since. In 1989 there were 354,693 part-time contracts, representing 1·6 per cent of male employment and 12·8 per cent of female employment. They may be for up to two-thirds of normal working time, defined as hours per day, hours per week, days per week or days per month. They may be either permanent or for a fixed term. Part-time employees have the same statutory rights as full-time employees, including representation rights.

There is no standard written form for permanent contracts, but most atypical contracts are required to be agreed on standard model forms issued by the employment offices, for registration there. The official form for a temporary job creation contract, for example, asks for the following information: name, age, national identity number, position, employer's name and address, tax and social security numbers, nature of employment, address, the number of employees on the payroll and in the workplace, the employee's educational level, and unemployment registration details. The employer is asked to state that in the previous twelve months he has not issued any unlawful individual redundancies or collective redundancies, that the post in question has not previously been filled by a three-year temporary job creation contract, and that the employee has not previously been employed on another three-year temporary contract by the company. The contract then specifies the nature of the work, the job grade, the hours, the pay and its component parts, annual leave entitlement, the probationary period, the length of the contract, with dates, and the compensation due to the employee on termination of the contract.

Controversy has recently arisen over legislation which took effect on 8 February 1991, obliging employers to forward to workers' representatives a copy of all new employment contracts for them to monitor that the legislation on temporary contracts is being observed. The government's reasoning behind this step is that an estimated 173,800 temporary contracts had exceeded their lawful limit in early 1991, yet employees were still being employed on those contracts. The trade unions have maintained for some years that there is considerable evasion by employers in this respect. The employers' confederation, the CEOE, has lodged an appeal against the ruling with the constitutional tribunal, and the result of the appeal is pending.

Rights and obligations

The 1944 law on employment contracts still applies to a number of aspects of employment, although most employee obligations are now detailed in the Workers' Statute. Article 20, for example, requires the employee to comply with the employer's instructions, provided the terms of the contract or a collective agreement or established custom and practice are not infringed. The same article also requires 'good faith' on both sides and requires the employer to give due consideration to 'human dignity'. Article 41 does allow the employee to refuse to carry out duties which 'substantially modify the terms of the employment contract' and allows an appeal to the labour inspectorate if the employee feels the right is being infringed. If the appeal is upheld, the employee may have the right to request termination of contract, with compensation of forty-five days' pay per year of service up to a maximum of forty-two months' pay. Nevertheless, although an employee can normally be required to work only to the terms of the contract and current legislation, 'occasionally, in urgent need to prevent imminent damage, or to cope with accidents, the employee will have to carry out work other than that in the contract — and be duly remunerated for it'.

The Workers' Statute gives employees a number of guarantees. Under article 4, for example, they are entitled to 'effective occupation, promotion and training . . . in respect of personal privacy and dignity, including protection from verbal, physical or sexual abuse, and the right to receive punctual salary payment'. In return, article 5 requires the employee to carry out the employer's instructions and to 'contribute to improvements in productivity'. Article 10 allows for an employer to contract with a group of employees. In this case the group leader assumes on behalf of the whole group the rights and duties that statute provides.

Employee inventions

According to the 1944 law on employment contracts, any inventions devised by the employee working under the terms of an employment contract belong to the employer. The employee is not entitled to any extra compensation for them, except when his or her own contribution to the discovery is so exceptional, or the invention so important, that these factors exceed the normal suppositions of the employment contract. Inventions made by the employee in other circumstances remain the employee's own property, except when the invention is due primarily to knowledge acquired through work for the employer, in which case the employer has rights of ownership or preferential use. In such cases, when the employer claims ownership of the invention, the employee is entitled to compensation relevant to the commercial worth of the invention. Any employee claim to benefit from an invention must be registered within three months.

Non-competition provisions

Article 21 of the Workers' Statute prohibits simultaneous employment with more than one employer when the work in question might be deemed to be unfair competition, or when one employer wishes specifically to forbid it in the employment contract. The employer may extend the prohibition for a period of up to two years after termination of contract for technical staff and up to six months for manual staff, provided that there is genuine competition from the other employer, and provided that the employee is adequately remunerated for it. If the employee is specifically remunerated to abide by such an agreement, he or she may withdraw from the agreement in writing, with thirty days' notice to the employer, forgoing any special payment which might have been agreed. Article 5 of the Workers' Statute also prohibits the employee from competing with the current employer.

Equally, when an employee receives specific technical training at the employer's expense, there may be written agreement reached by the two sides that the employee will remain in employment with the employer for a certain period of time – which may not exceed two years. If the employee breaks this agreement by leaving within the agreed time, the employer is entitled to seek compensation and damages. Collective agreements sometimes address this issue: for example, Unión Explosivos Río Tinto obliges any employee trained at the company's expense to remain with the company for a further two years — or compensate the company for the cost of the training. Moreover the employee is required to give six months' notice of termination of contract if wishing to leave before the two-year period has elapsed.

Confidentiality

Infringing company confidentiality is regularly defined as a 'very serious breach of contract', as for example in the Iberduero agreement and national perfumery and chemicals agreement, which also applies the same definition to 'taking part in activity contrary to the interests of the company'. Sanctions render the employee liable to suspension and fines of between twenty and ninety days' pay. In the national perfumery and chemicals agreement, sanctions range up to dismissal.

Discipline

The Workers' Statute prescribes procedures for disciplinary action against individual employees, by which management is empowered to impose sanctions on an employee for breach of discipline. The Workers' Statute outlines breaches of conduct which warrant dismissal: they include repeated unjustified absence or unpunctuality, indiscipline or disobedience, verbal or physical abuse of employers, colleagues or their families, breach of faith or confidence, poor working habits and repeated alcohol or drug abuse.

The detailed definition of breaches of discipline and the appropriate sanctions occurs either in surviving labour ordinances or collective agreements. A model agreement in this respect is the national chemical agreement. It defines breaches as being light, serious or very serious. Light breaches include late arrivals at work; failure to register sick; leaving the workplace; neglect of working materials; lack of hygiene or cleanliness which impedes production; poor service to the customer; failure to notify a change of address; workplace conversations on extraneous affairs; unjustified absence of one day per month. Serious breaches include: more frequent unpunctuality; unjustified absence of two days per month; feigning illness or accident; disobedience towards managers; negligence; repetition of a light breach within the same quarter, after written warning. Very serious breaches are: more than ten late arrivals in a six-month period or twenty in a year; fraud or theft; breaching confidentiality rules; verbal abuse of management or colleagues or their families; negligence leading to a serious accident. In the event of these breaches of discipline occurring, management is empowered to take action, while being required to inform workers' representatives at the same time. Maximum sanctions are as follows: for a light breach, verbal or written warning and suspension of employment and pay for up to two days; for a serious breach, suspension of employment and pay for three to fifteen days; for a very serious breach, from sixteen to sixty days' suspension of employment and pay up to termination of contract (see below, termination of contract).

Where employees have a grievance or object to sanctions imposed by an employer — and especially in cases leading to dismissal or suspension — they may take their case to the local office of the Institute of Mediation, Conciliation and Arbitration (IMAC). Once a part of the Ministry of Labour, IMAC is gradually being transferred to the governments of the Autonomous Communities. By 1990 this had happened in seven of the seventeen communities. In 1989 IMAC handled a total of 452,186 cases.

Working time

Hours of work

By statute the maximum working week is forty hours. If the working day is not split (a common practice in Spain) there must be a fifteen-minute break, which may or may not be paid. A normal working day must not exceed nine hours' actual working time, and there must be a minimum twelve hours' break between one working day and the next.

Most collective agreements now define working time in terms of annual hours. In 1989 the average working time collectively agreed was 1,771 hours (an average thirty-nine hour week). The typical figure for company agreements was 1,745 hours per annum. The average working year has fallen by some seventy-five hours since 1983. Current policy by the major trade unions, UGT and CC.OO, is to press for a thirty-eight-hour week.

One feature which distinguishes Spain from most European countries is the persistence of the *siesta*. Many companies, especially in the office, service and retailing sectors, have a working day of 08.00 or 08.30 to 14.00, 14.30 or 15.00. Others work 09.00 to 14.00 and 17.00 to 20.00. This is at least in part explained by the climatic conditions of the summer months, and there has been a generalized resistance to any move to continuous day working for that reason.

Special working hours for certain employees

Employees in certain occupations have their hours of work regulated by separate statute. This allows for more flexible regulations to cover occupations whose working conditions so require them. The following groups of employees are affected in this way:

Watchmen, caretakers, residential security staff must have a minimum one hour meal break, a daily rest entitlement of at least four hours, and a break of at least ten hours between shifts. Residential security staff may be required to work up to twelve hours daily, with entitlement to four hours' rest, including mealtimes, per day.

Road transport. Long-distance drivers must not drive more than four hours without a break. Breaks must be a minimum thirty minutes, distributable throughout the driving period. Normally drivers must not work for more than nine hours daily.

Employees working in remote locations, such as radio transmitters or isolated railway junctions may have their working time calculated in fortnightly periods, with appropriate time off in lieu for work beyond forty hours per week.

Limitations on working time in conditions of risk. When working conditions are exceptionally dangerous, arduous, unhealthy or toxic, working time may be limited, either by agreement or by the intervention of the labour inspectorate, with no reduction in pay.

Flexible working hours

Flexible working time arrangements are not very widespread in Spain. However, collective agreements do provide some examples. Repsol, for example, operates a scheme whereby employees at corporate headquarters in Madrid have a flexi-time scheme with winter and summer variations. In winter forty hours are worked, with two core periods: 09.30–13.30 and 15.30–17.00. Employees may start from 08.00–09.30 and leave between 17.00 and 19.30. In summer (1 July–15 September) thirty-five hours are worked, with a core period of 09.00–14.30. Starting times may be between 07.30 and 09.00, leaving times between 14.30 and 15.00 or 15.30 to 18.30. Zurich Insurance operate a normal day of 07.30–15.00 and allows thirty minutes' flexibility in arrival and departure times. Alternatively, employees may work 08.00–13.30 and 15.00–17.00.

Overtime

By statute, overtime payments (for time worked beyond the maximum forty hours per week) must be at a rate of at least 175 per cent of normal time, whether working on a weekday, weekend or public holiday. Statute also limits the amount of overtime that may be worked: no employee may work more than eighty hours' overtime per annum. Any overtime worked must be recorded and totalled. Overtime is prohibited on night shift. Many collective agreements specifically charge the works committee with monitoring these conditions.

Since the time of the Acuerdo Marco Interconfederal (AMI) national framework agreement between the employers' confederation CEOE and the socialist-led trade union UGT in 1980, trade unions and employers' organizations have agreed to reduce and, if possible, eliminate overtime working in the interests of job creation. Collective agreements therefore contain a commitment to limit overtime. The AMI also established the distinction between overtime as an extension of normal company activity — which it agreed to eliminate — and 'structural overtime', caused either by *force majeure*, involving a serious threat to the company, sudden peaks in demand which cannot be met by other means, or unavoidable urgent requirements of production. Such 'structural overtime' is tolerated.

As a further deterrent to overtime, in 1981 the government ruled that overtime pay should bear a higher social security rate of contribution when it is for other than 'structural' purposes. In 1990 the structural overtime rate was 12 per cent for the employer and 2 per cent for the employee; other overtime is levied at 24 per cent for the employer and 4·8 per cent for the employee.

Statute also allows time off in lieu to be negotiated instead of overtime pay. Such time off must be at a rate of 175 per cent of normal time.

Night work

Night work is forbidden by statute to those under 18. Statute also requires extra payments of at least 25 per cent of basic pay for work between 22.00 and 06.00 hours. Collective agreements sometimes improve on this. For example, Cedosa gives 30 per cent extra, Ford gives 40 per cent extra on basic, and Renault reduces the night shift to four hours on Saturday and Sunday nights, paid as if eight hours were worked.

Time off

Annual leave

Statutory minimum leave is thirty calendar days' (four weeks') paid annual leave (twenty-two working days), which may be divided into two periods and may not be replaced by pay. Most employees take their annual leave in August, and many companies close down for the whole month for that reason. There is a small but growing trend towards taking part of the annual leave in the Christmas and New Year period. Employers have a statutory right to exclude peak production periods from holidays, after consulting staff representatives. Employers are also obliged to establish an annual holiday rota, by agreement with staff representatives: employees with children enjoy statutory preference as regards holidays during school holiday periods. The employer must display the holiday rota, giving employees at least two months' notice of holiday periods.

Some collective agreements improve on these rights. For example, Citroën Hispania, whilst stipulating that holidays must be taken in July and August, or exceptionally in June or September, provides one extra day for each three years' service, up to a maximum of seven days. Employees with this right will take thirty-one calendar days' uninterrupted vacation, and the remainder outside the months of July and August. Employees with less than one year's service receive holidays proportional to their service, with calculations of service based on a calendar of 1 January to 31 December in Barcelona and 1 July to 30 June in the other centres.

Public holidays

All employees are entitled to a minimum of fourteen days' paid public holidays per year in addition to paid annual leave. Each December the government sets the calendar of public holidays for the following year by announcing twelve dates which apply to the whole country, and allowing local authorities to allocate another two days for local celebrations (fiestas) which are usually connected with the local patron saint of a village or town. Of the twelve dates announced by central government, three may be replaced by others of local importance by the governments of the autonomous communities.

The twelve days which are most often national public holidays are: 6 January (Epiphany), Holy Thursday, Good Friday, May Day, Corpus Christi, 25 July (the feast of St James, patron saint of Spain), 15 August (the Assumption), 12 October (feast of the Virgin del Pilar, patroness of Spain and Day of the Hispanic People), 1 November (All Saints), 6 December (Constitution Day), 8 December (Immaculate Conception) and 25 December.

Other dates often used for the other two public holidays are 1 January, 19 March, Ascension Day, 29 June, 31 December, 15 May (Madrid), 26 December (Barcelona) and 31 July (Bilbão). The government has statutory powers to move a public holiday to a Monday if it considers it in the best interests of working arrangements. When a public holiday falls during annual leave time or on a weekend the employer does not have to compensate.

When a public holiday falls on a Tuesday or a Thursday it is widespread practice for employees to take the Monday or Friday as part of annual leave, a practice known as *puente* (bridging).

Paid leave of absence

Provided due notice and good reason are given, employees are entitled by statute to the following paid leave:

- Marriage: a fortnight.
- Birth of a child or death or serious illness of a close member of the family: two to four days.
- Public or unavoidable duty: whatever time is necessary. If it amounts to more than 20 per cent of working time in a three-month period the employer is entitled to declare the employee laid off, with the right of return to the post.
- To fulfil elected union duties.
- Nursing mothers may have one hour off per day, divided into two periods, for feeding children up to nine months. Alternatively they may shorten their working day by half an hour.
- Time to attend examinations connected with education and training.

Unpaid leave of absence

After one year's service, employees have a statutory right to request leave of absence, with preference for reinstatement, to look after a child for up to three years.

Employees are entitled to two to five years' leave of absence for any purpose, with no preferential reinstatement, after one year's service. There must be a minimum four years between each leave period.

Employees whose public-service duties or senior trade union work requires them to be absent for more than 20 per cent of working time in a three-month period may be required by the employer to take temporary unpaid leave of absence, with time off counting towards increments and their post guaranteed, provided they request reinstatement within one month after their duty ends.

Collective agreements sometimes improve on these provisions. For example, the chemical industry agreement allows employees with six months' service to request unpaid leave of up to three months. The employer is required to respond favourably when production circumstances permit, and when no more than 2 per cent of the work force will be affected; two years must elapse before the next such leave.

Maternity leave

There is a statutory right to sixteen weeks' maternity leave (or eighteen weeks in cases of multiple birth) on a state benefit at 75 per cent of previous earnings. At least six weeks of this leave must be post-natal. In the event of a mother becoming seriously ill or dying as a result of the delivery, the father is entitled to the six weeks' post-natal leave. In any event the father is entitled to two days' paid leave for the birth.

Either the father or the mother has a statutory right to up to three years' unpaid leave to care for a child, or to reduce working time between 33 per cent and 50 per cent to care for a child under 6, with a corresponding reduction in pay. Mothers also have a statutory right to an hour's absence from work daily to feed a child up to nine months old. This may be divided into two periods: alternatively the mother may shorten her working day by half an hour.

Collective agreements sometimes improve on this; for example, at Telefónica either the father or the mother may take time off to feed the child if both are employees of the company. Citröen Hispania allows a further thirty minutes per day unpaid leave for feeding the child.

Paid time off for trade union/worker representatives' duties

Elected works committee members or workers' representatives have a statutory right to paid leave to carry out their duties. The amount of leave is based on a scale according to the size of the company, as follows:

- 0–100 employees: fifteen hours per month per representative.
- 101–250 employees: twenty hours per month per representative.
- 251–500 employees: thirty hours per month per representative.
- 501–750 employees: thirty-five hours per month per representative.
- 751+ employees: forty hours per month per representative.

By collective agreement these time credits may be consolidated to allow representatives to be full-time workers' representatives with no loss of salary or entitlement. The chemical industry agreement also stipulates that time spent in collective bargaining negotiations will not count as part of the maximum permitted time off. It also allows for time spent in trade union training activities to be accepted as qualifying for part of the statutory time allowance. The agreement at Telefónica allows for up to twenty hours per year for all employees to attend union branch meetings (meetings of up to twenty minutes each), and grants an extra allowance of eighty hours per month for trade union regional officers where branches have at least 25 per cent affiliation of the work force.

Military service

When employees are called up for military service, or volunteer, their job must be reserved for them, provided they take up employment again within thirty days of termination of service.

Retirement

The maximum permitted age for employment is 69, but the retirement age for state pensions is 65. The state retirement pension is relatively generous, and is earnings-related. With a minimum fifteen years' contributions a retiring employee receives a minimum 50 per cent of covered earnings, up to a current maximum state pension benefit of Pta 203,280 (£1,117) × 14 per annum. A full pension is earned after thirty-five years' contributions. Benefit levels are expressed as a percentage of the social security calculation base, subject to an absolute maximum, and vary according to occupational category; there are twelve categories in all.

Under a job creation scheme *early retirement* on full pension is permitted at 64, provided the employer recruits a replacement employee from among the registered unemployed. Under a further job creation measure, the 'relief contract' (*contrato de relevo*), employees may take partial retirement from 62, with full pro rata pension entitlement, provided the employer recruits a replacement employee from the registered unemployed, on a part-time contract.

Agreed company schemes often improve on statutory requirements; one of the most generous is at Repsol, which tops up the state pension to 100 per cent of the employee's basic final pay. There are supplementary state pensions available to those aged over 67 lacking sufficient contributions to be entitled to a retirement pension.

Private pension schemes are regulated by the 1988 legislation *regulación de los planes y fondos de pensiones*. The law provides for tax relief on approved funds and employer/employee contributions, provided legal requirements are met: these cover aspects such as financing arrangements and the distribution of assets, actuarial review, and the management structure of funds. Three types of scheme are permitted under the law:

- Company schemes financed by employer/employee contributions.
- Schemes for members of professional associations or trade unions.
- Individual schemes via financial institutions.

Schemes must be funded, and the management of approved pension funds is by boards composed of fund directors, participants and beneficiaries, with the theoretical possibility of employee control over fund investments.

Equal treatment

Article 14 of the constitution declares Spaniards equal before the law, regardless of birth, race, sex, religion, opinion or any other social or personal condition or circumstance. Article 35 guarantees the right to free choice of profession, equality for promotion, and appropriate remuneration regardless of gender. Article 4 of the Workers' Statute further prohibits discrimination on grounds of gender, civil status, age, race, social position, religion, politics, membership or not of a trade union, and linguistic background. It also prohibits discrimination on the grounds of physical, psychiatric or sensory disability, provided the employee is capable of carrying out the work in question. Article 17 reinforces these conditions by prohibiting collective or individual agreements from discriminating on the above grounds, or on grounds of family relationship with other employees. Article 28 reiterates that there must be equal pay for equal work by men and women.

Spain has also ratified EC and ILO directives on equal pay and non-discrimination. The Instituto de la Mujer (Institute for Women), part of the Ministry of Social Affairs, has investigated the actual employment experience of women in Spain, where the number of women in the work force rose from 3,752,200 in 1981 to 5,141,300 in 1989. On average, women receive some 20 per cent less pay than men, largely through being concentrated in lower-paid occupations or posts. The government has responded by creating a special employment contract to boost the employment of women in professions or activities in which they are underrepresented: this provides a grant of up to Pta 250,000 (£1,375) — or Pta 500,000 if European Social Fund money is available — to employers who offer a permanent contract to unemployed women under 25 or long-term unemployed women over 25. Take-up of this subsidy has been slow — in 1989, for example, only 186 women obtained employment under the scheme.

Some collective agreements make provision in the field of equal opportunity. The chemical industry national agreement establishes a joint standing committee to study the effects of discrimination on women in the industry. The engineering (small firms) national agreement pledges equal opportunity in employment, training, promotion and career development. It also proposes that job descriptions, the organization of work and working conditions should be designed to offer equal access to men and women, and establishes a joint standing committee to promote these objectives. At Repsol the agreement stipulates that women must have equal opportunities for employment, training, promotion and career development, and supports positive action to achieve this. The committee is charged with analysing the present situation, taking steps to increase the presence of women in traditionally male jobs, and developing special training programmes for women, particularly in IT administration, technical training and management training.

Work environment

Health and safety are regulated by legislation dating from 1971. The 1978 constitution charged the government with the duty of controlling health and safety, and draft legislation is being considered to revamp the whole health and safety structure in Spain. The labour inspectorate (*Inspección de Trabajo*) monitors conditions, and has issued twenty-nine codes of practice for different sectors of industry.

The Workers' Statute stipulates statutory health and safety measures in the workplace, and decree 432/1971 details the implementation of the structure still in force. It obliges all companies and workplaces with more than 100 employees (or fewer when 'specially dangerous activities' are concerned) to establish a health and safety committee, consisting of a chairman appointed by the employer, a qualified health and safety technician and a qualified medical specialist, a qualified first-aid member of staff, the head of company security, a number of elected staff representatives (three in companies under 500 employees, four in companies of 500–1,000 employees, five in companies of more than 1,000 employees) and a secretary appointed by the company from among the white-collar staff.

Should three-quarters of elected staff representatives consider there is a grave risk of accident due to the prevailing working conditions, the work force are entitled to stop work, informing the employer and the labour inspectorate, and receive a ruling from the labour inspectorate within twenty-four hours. Employers who infringe health and safety requirements are liable to a fine ranging from Pta 30,000 to Pta 300,000 (£165 to £1,650).

Collective agreements sometimes make detailed provision on health and safety. Unión Explosivos Río Tinto guarantees, among other measures, eight hours' health and safety training annually for all employees. Telefónica allows pregnant employees to request transfer to more suitable work. It also provides three ten-minute breaks per shift for VDU operators, when staff rotation is not possible, and annual medical checks for VDU operators, with the onus on the employer to pay for any prescribed remedies for complaints occasioned by constant VDU use (such as glasses or contact lenses).

Individual termination of contract

The Workers' Statute (law 8/1980, articles 49–56) provides for several circumstances in which the employment contract may be terminated:

- By mutual agreement. In that case a document known as a *finiquito*, detailing pay received and other employment details, is issued, and the contract is terminated, with no entitlement to compensation.
- By cause stipulated lawfully in the employment contract.
- On conclusion of the period of the contract or the job or service in question. In these cases, when there is a fixed-term contract of more than twelve months, the employer must give at least fifteen days' notice. Otherwise the contract is considered extended in-

definitely, except when the temporary nature of the job is self-evident.

- Through death, retirement or legal incapacity of the employer. In the case of the death or retirement of the employer, the employee has a statutory right to compensation. When the employer becomes legally incapacitated the employment authorities are required to open lay-off proceedings (*regulación de empleo*).
- Termination by the employee. In this case the employee simply has to observe notice requirements as stipulated by the contract or collective agreement.
- Owing to the death, total disability or serious invalidity of the employee. When the employee is declared permanently disabled as regards normal working duties the employer may terminate the contract or offer another appropriate job.
- Retirement of the employee. The contract may be terminated when the employee reaches the state retirement age of 65.
- *Force majeure*. In the event of unforeseen circumstances which interrupt the normal working of the business. The Statute outlines the procedures for confirmation of *force majeure* by a 'competent authority'.
- For economic or technological reasons. When a business is suffering losses or needs to restructure technologically to improve performance, there are statutory procedures for initiating redundancies. Two or more dismissals for economic reasons constitute a collective redundancy, involving consultation procedures.
- Failure of the employer to abide by the terms of the contract. In certain circumstances the employee has the right to request termination of contract, with compensation due as in the case of unlawful dismissal. This generally applies when the employer alters working conditions such that the employee suffers in career or training prospects, or when the employer is guilty of serious breach of contract, as in repeated delay or failure to pay wages. In these cases the employee may be entitled to compensation of forty-five days' pay per year of service, up to a maximum of forty-two months' pay.

Individual dismissal

The two main forms of individual dismissal provided for in the Workers' Statute are:

- Dismissal on disciplinary grounds.
- Termination for 'legally valid reasons of an objective nature'.

A *disciplinary* dismissal is justified on the following grounds (article 54):

- Repeated unjustified absenteeism or unpunctuality.
- Indiscipline or disobedience.
- Verbal or physical aggression towards the employer, colleagues or their families.
- Abuse of contractual good faith, or breach of confidence.
- 'Consistent and wilful reduction of normal or agreed output'.
- Habitual alcohol or drug abuse adversely affecting performance.

The employer must give the employee written notice of dismissal, citing the reasons, within sixty days of learning of the breach of discipline, and in any case within six months of the date of the offence. The notification must indicate when the contract will be terminated

Appeals and compensation payments

The employee has a right of appeal against dismissal. Should a court declare the dismissal unlawful, the employer has five days in which to decide whether to reinstate the employee or pay compensation. Reinstatement is mandatory only where the unlawfully dismissed employee is a work-force representative. Compensation is paid at a rate of forty-five days' pay per year of service, up to a maximum of forty-two months' pay.

Compensation is reduced by 25 per cent for small firms employing fewer than twenty-five workers, with 40 per cent of the outstanding amount paid by the Wages Guarantee Fund. In addition, any lost earnings between the date of dismissal and the court's ruling (or finding alternative employment) are payable. If the court takes longer than sixty days to give a ruling, any excess is covered by the state.

Dismissal *on objective grounds* is possible on four counts:

- The employee's lack of skill. There are sufficient grounds for dismissal if the employee's inadequacy to the task becomes apparent after the probationary period is over.

- The employee's inability to adapt to technical change. If, after two months' following the introduction of a technical change, an employee proves unable to adapt to reasonable new demands, the contract may be suspended for up to three months for retraining and adaptation purposes before proceeding to dismissal.
- The need to reduce costs by eliminating a post. This is permissible in companies with fewer than fifty employees, provided the employer can prove economic necessity. If a post exists in another workplace, or if the same post is reopened within one year, the dismissed employee has preferential entitlement to the vacancy.
- Absence from work. A certain rate of unjustified absenteeism is sufficient ground for dismissal. It must amount to more than 20 per cent of working days in a period of two consecutive months, or more than 25 per cent in any four non-consecutive months in any period of twelve months. Absence here excludes: lawful strike, certified illness, statutory time off, agreed unpaid leave or annual holiday.

In all these cases the employer must give the employee written notice, citing the reasons for dismissal. The period of notice must be at least one month for employees with less than one year's service, two months for between one and two years' service, and three months for two or more years' service.

Severance payment for objective dismissal

The employee is entitled to compensation of twenty days' pay per year of service, up to a maximum of twelve months' pay, which must be made available to the employee at the time of serving notice. The employee is also entitled to six hours' paid absence weekly to seek new employment.

Termination of executive contracts

An executive wishing to resign must give three months' notice of termination, and may have the right to compensation in certain circumstances, such as substantial changes in working conditions, delay in receiving pay, failure by the employer to fulfil contractual obligations, or change of ownership of the company with substantial repercussions on personnel or business activity. In such cases the executive may be entitled to seven days' pay per year of service, up to six months' pay.

If the employer wishes to terminate the executive contract he must do so in writing with three months' notice, with compensation of seven days' pay per year of service, up to six months' pay. If the employer dismisses the executive for failure to fulfil contractual obligations, he is liable to pay compensation of twenty days' pay per year of service, up to twelve months' pay, should the dismissal be ruled unlawful.

Appendix

Ministry of Labour and Social Security (Ministerio de Trabajo y Seguridad Social)
Augustín de Bethencourt 4
28003 Madrid
tel. +34 1 253 6000/253 7600
fax +34 1 233 2996

Instituto Nacional de Empleo (INEM)
Calle Condesa de Venadito 9
28027 Madrid
tel. +34 1 585 9888
fax +34 1 268 3981/268 3982

Asociación Española de Directores de Personal (AEDIPE), the Spanish Association of Personnel Directors
Moreto 10
28010 Madrid
tel. +34 1 468 2217

Confederación Española de Organizaciones Empresariales (CEOE), the Spanish employer's organization
Diego de León 50
28006 Madrid
tel. +34 1 563 9641
fax +34 1 262 8023

Instituto de la Mujer (equal opportunities agency)
Almagro 36
28004 Madrid
tel. +34 1 410 5112

British Chamber of Commerce in Spain
Paseo de Gracia 11A
08007 Barcelona
tel. +34 3 317 3220
fax +34 3 302 4896
and
Pl. Santa Barbara 10–1
28004 Madrid
tel. +34 1 410 7064
fax +34 1 410 4605

Further reading

INCOMES DATA SERVICES *European Report*, published monthly, provides news updates and features on developments in the areas of collective bargaining, employment law, state benefits, and European Community law. The *International Documents* series consists of more in-depth reference documents on a country-by-country basis. Updated annually, the series consists of one set of Documents covering Pay and Conditions, and a further set on Commencement and Termination of Employment. Further details of IDS's International Service are set out on p. xvii.

INTERNATIONAL LABOUR OFFICE (ILO). The ILO publishes a series of publications termed *Conditions of Work Digest*. Two issues appear annually, each focusing on a single topic. The publications typically outline legal principles with examples of practice, and the views of trades unions, employers' organizations and government.

Recent issues have covered telework, childcare, and new working schedules.

The ILO also publishes an *International Directory of Occupational Safety and Health Institutions* (1990).

Details from the ILO London office: ILO, Vincent House, Vincent Square, London SW1. Tel. 071-828 6401.